Hardrock Man

Whispers from Underground

Oral History of Mining
"The World's Greatest Gold Camp"

by
Sylece Andromeda

Hardrock Man

Whispers from Underground

Oral History of Mining
"The World's Greatest Gold Camp"

Hardrock Man - *Whispers from Underground*
Oral History of Mining "The World's Greatest Gold Camp"

Title page photo courtesy Ed Hunter and Gene Leaf; pages 87 & 264, Jim McGee, Sr., page 73, Ed Hunter; page 245 Ted Johnson; District photo section includes Hunter, Johnson & Price Collections. Other photography by the author. T. Ronan and W. Roy recorded 1975 by Harold M. Hern, donated 1999 to Sylece Andromeda.

We don't each remember things in the same way, and sometimes what we think we remember isn't really what happened... or is it? Therefore, let it be understood that names, dates, events, opinions, proper spellings and everything else contained herein may or may not be but a figment of the speaker, reader, or author's imagination. Then again, maybe it happened just like they say. One never knows.

Published by: Poverty Gulch Publishing, PO Box 598 in Cripple Creek, CO 80813. Printed and bound in the United States of America.

ISBN: 978-0-9787747-1-4 *13-digit*
0-9787747-1-X *10-digit*

Library of Congress Catalog Number: 2007925559

First Edition Perfectbound *Limited First Printing* *April, 2007* 250 copies

3-2-1

Ready to Shoot

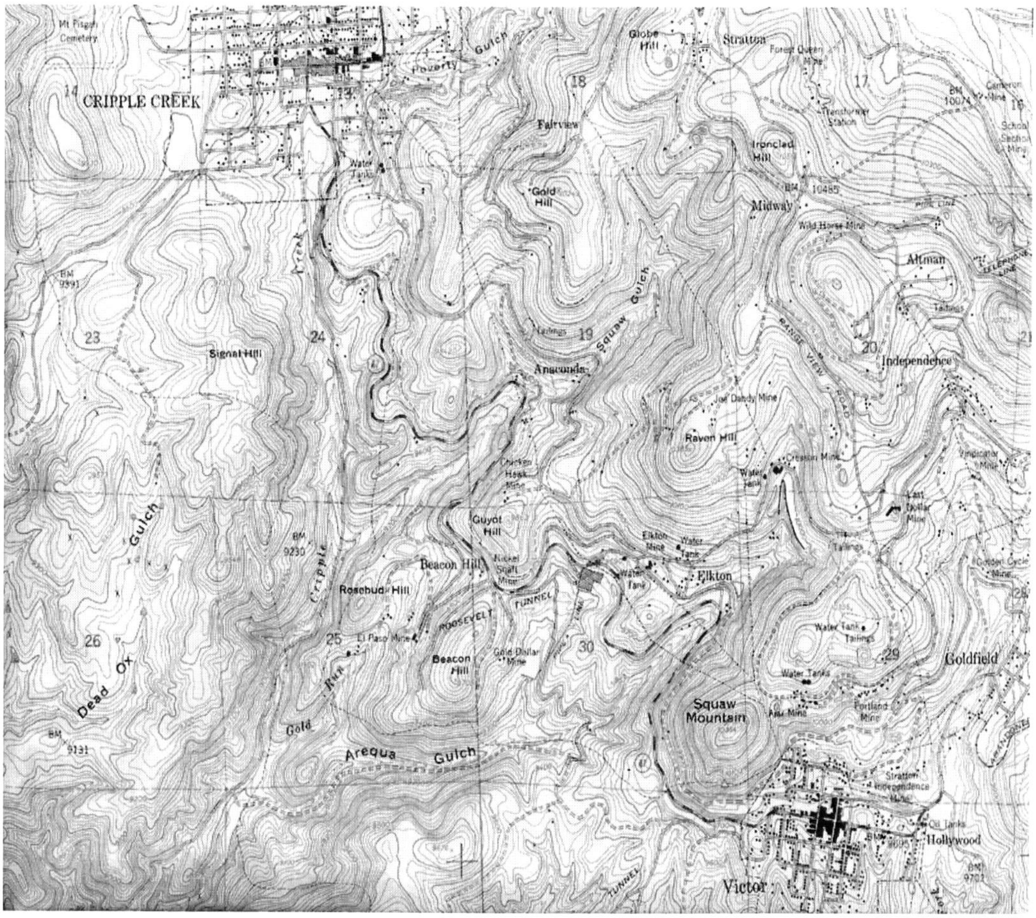

United States Department of the Interior Geological Survey
Cripple Creek South Quadrangle Topographical Map 1951 rev. 1975

Contents

1. Introduction 1
2. Prologue & Secret Mine Camp Words 3
3. Teasers 7
4. Mule Work on the Bull Gang Again 11
5. Hunting Season's a Good Excuse to Say Goodbye 16
6. Communication, Hairnets & White Boots 39
7. The Skipper, The Mule Skinner and The Dump Rat 60
8. You're 'Da Man 73
9. Well, It Can't Go That Far 84
10. Mining is a Good Way to Go in the Hole 87
11. Commander Drum, Beaver & Young Men End the Era 96
12. Beans on the Table 121
13. Channel 9 News 143
14. Riding the Rim 150
15. The Mad Mucker 161
16. Widow Makers & Hand Cars 183
17. Tour Guide as Timber Man 189
18. Thanks a Lot, Bastard 197
19. Tramp Miner to Mine Owner 212
20. The Muckin' Machine's Hittin' the Boss in the Ass 223
21. How to Buy Your First Car 235
22. Every Job But Boss 241
23. Charmed, I'm Sure 245
24. High-Grading vs. Pocket Rock 264
25. Glossary 288

Index of Photographs

Page	Location
Title page	Underground in the Ajax Mine
11	Anchoria Leland Ore House
16	Grace Greenwood Mine
39	Calaverite specimen Bobtail Vein, Ajax Mine
60	Theresa Mine
73	Colorado State Code of Mine Bells
84	Ajax Mine
87	El Paso Mine
96	International Mine Conveyance
121	Independence Mine
143	Hull City Mine
150	Carlton Tunnel Portal
161	Anchoria Leland Hoist
183	Colorado Gold King Mine
189	School Section Mine Ore House
197	Hoosier Mine
212	Gold Band Mine
223	International Mine
235	ABKAP Mining
241	Mary Nevin Mine Ore House
245	Mary Nevin
264	Cresson Mine Collapse
288	Vindicator Mine
294	Anchoria Leland Mine

Cripple Creek - Victor District Photographs

Ajax

Anchoria Leland

El Paso- double-drum hoist

Cameron

Hiawatha

Sylvanite

Cripple Creek - Victor District Photographs

El Paso

Mary McKinney

Mary Nevin

Portland #2

Elkton

Ajax Ore House

Cripple Creek - Victor District Photographs

Empire Lee

New Last Dollar

Mucking the Mary Nevin

12B Eimco Mucking Machine

Gold King

Deadwood

ABKAP (adit)

Cripple Creek - Victor District Photographs

Rails of the Cresson

Taking out Midland Terminal Railroad Track

Independence Mill Remains

Carlton Tunnel

Undated message marked on rock using a carbide light Mary Nevin 270' Level

Cripple Creek - Victor District Photographs

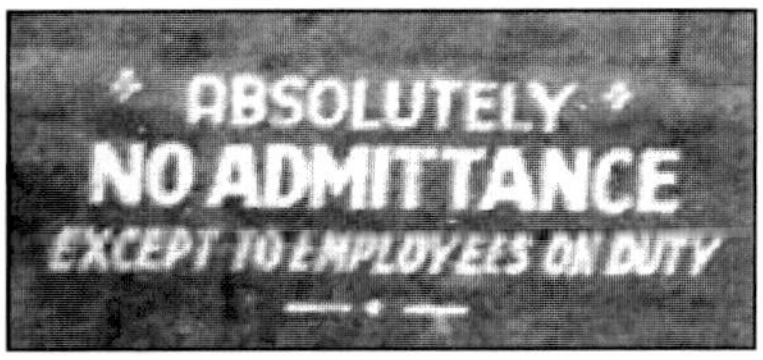

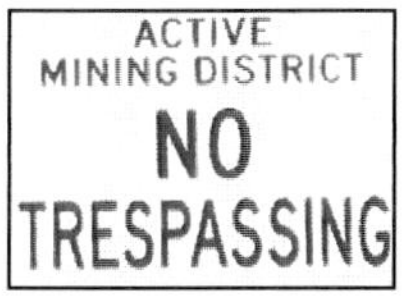

Powder Shack Sign:
DRIVE NO NAILS
Use Wooden Pegs

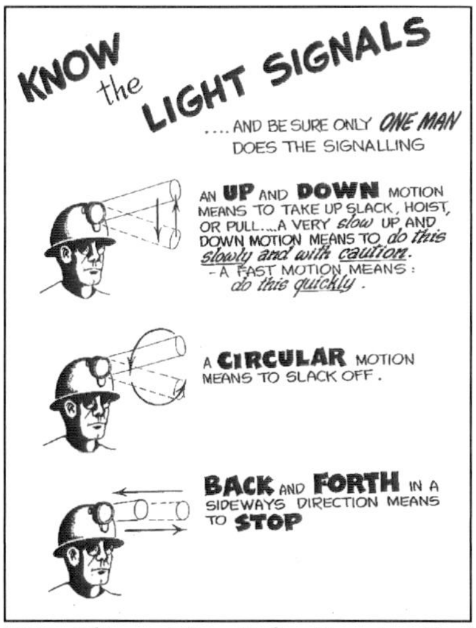

1979 Climax Mine Safety Practices Book

Cripple Creek - Victor District Photographs

Ajax erecting new frame around the old

Victor G.M. Co.

Isabella

Delmonico (Maybe)

Pinto Shaft

Prince Albert

Introduction

A world today without metals and minerals is hard to imagine. Unfortunately, very little has been written about the people who got those needed materials out of the ground to make everything from our houses to our computers.

The first publication devoted to getting metals out of the ground was the book "De Re Metallica" written by Georgius Agricola in the 1500s. This was a very thorough view of the industry at that time, but concentrated on processes.

Many of the books on mining today are technical publications, except the numerous volumes written about various gold rushes. However, like most of the books that followed Agricola, the accounts are faceless, for very few concentrated on the people involved and those that did were written most often in the words of the author and his or her perceptions.

The years following World War II to the present were one of great advancements in equipment and techniques used in mineral extraction, little of which is recorded in the words of the people who did the work.

Sylece Andromeda has reversed this trend by recording oral interviews with workers from the mines underground in the great Cripple Creek - Victor Gold Mining District of Colorado. She has chosen her interviewees to cover a cross-section of the main jobs at the mine from the miners themselves, a hoistman and an owner/manager to cite only three examples in the collection.

With a knack for drawing interesting and humorous incidents from her interviewees, Sylece puts a face on those miners for the reader who has never been underground. To the miner who has been there before, these tales will bring back many memories. The author has been working against time as many of those who worked the small underground mines have passed on.

The trend in mining has been to large scale operations, utilizing huge rubber-tired trucks and loaders. This has reduced the intensive physical labor previously experienced in mining underground and provided the worker with far safer working conditions. This larger scale of operations has also permitted the mining of lower and lower grades of material, thereby conserving our natural resources. Yet mining has lost the comradeship of those working in close, dark and dangerous quarters underground.

Sylece's oral interviews have preserved an important phase in the life of "The World's Greatest Gold Camp." The miners and others featured in her book are among the last of those who mined over 21,000,000 ounces of gold from underground in this District.

Today, surface mining is producing more millions of ounces of gold and it is hoped that the author will continue her excellent work by preserving the words of the surface miners of today, for tomorrow's generations.

Edgar T. Hunter, E.M.
May 5, 2005

Prologue

Raised in "The World's Greatest Gold Camp," it was a natural to be fascinated with underground mining, which put food on the family table. Dads, brothers and friends disappeared to the underground each day or night for work, returning topside with 'pretty rocks,' and other-worldly tales. Well, mostly they returned; sometimes they didn't. These are the stories of both kinds. Here the underground comes to life with mystery, tragedy and excitement through the voices of those who mined there.

Many have asked what it was like to collect stories from such colorful characters. Well, let me tell you: Miners really are ornery critters, and they are a lot of fun. Their stories amazed, entertained, horrified and yes, sometimes grossed me out. It was a great experience, and a real education.

One interview found me sitting cozy on a little couch with a growling dog and his old master who would periodically smile and say, "Oh my, he doesn't usually take to strangers so well!"

With concern that I might later get thirsty, one loaded my pockets with soda pops. I was offered a vegetarian picnic lunch, sang hymns to and drank beer after shift with more than a few. Some blushed above their beards when they realized I knew their term 'piss ditch,' while others showed off rare telluride gold-ore specimens handed down through the generations, or that they themselves had mined. Some learned I like hot tea, and made a point of fixing it special. Work-hardened hands serving dainty tea cups. That is a sight to see.

Such kindness and consideration. These miners may be ornery, but they have hearts of gold. Some I have known all my life, others I had never before met. All have been endeared to me through the gift of his words and stories, and it is my pleasure to know them.

Much appreciation to each of these guys, as well as many others in the District for their help and shared enthusiasm. Special thanks to Mr. Jim McGee, Sr. and Mr. L.R. Myers for answering my very many miscellaneous questions through the years; likewise to the gracious Mr. Ed Hunter, who became a dear friend and mentor; and to Mr. Harold Hern, who donated interviews he recorded when I was just in kindergarten. Rough recordings that had never been heard, they really were just waiting for the right moment to arrive. Timing is everything.

Speaking of which, it's that time....

Grab your sense of adventure, pull on your mucking boots and come on down underground. It's a lovely day for a tour.

Sylece Andromeda
2005

Psssttt... Secret Words of the Mine Camp

Why it's called a Gallas Frame, of course!

Most "Old Timers" in the Cripple Creek - Victor Gold Mining District voice an identical word, pronunciation and phonetic spelling when asked to "Name the structure that sits above the shaft" = = = *GALLAS FRAME.* Though the word is of English origin and might 'properly' be spelled Gallows Frame, that isn't what we called them, so it would be misrepresentative to pretend that we did. We call them Gallas.

Headframe is a term that was commonly used throughout the world by engineers, prospectors, tramp miners and writers, amongst others. Yet to the community here, they were gallas frames. Then folks passed on, or moved away. Several felt their use of gallas displayed a level of 'local ignorance' so they switched to using headframe. It would appear that outside influences in the late 1970s by investors coming to the area had a lot to do with 'headframe' finally overtaking as the prominently accepted and expected term.

Just like the frames they represent, it appears "*Gallas*" will fade away and be forgotten... but now you know.

Another 'local word secret' is the pronunciation of a rare telluride ore for which this District is famous: 'Calaverite.' It was named for Calaveras County, California, where it was first discovered. So, the word is properly pronounced just like the county name: "Cal-a-VER-ite."

But not in this District! In this mine camp, we say: "Ca-LAV-er-ite."

Why did the District make up their own word pronunciations? There were a lot of people through here at one time. Maybe local vernacular was one way to tell who was new, and who'd been around and could maybe be trusted.

Miners think of "a round" as a blast, and "a blast" doesn't mean having a good time. The ceiling of the tunnel where they work is called the "back," and "high-grade" can be a noun, adjective or verb. How confusing! After all that, the last thing you'd want to explain is that you aren't a hangman at the Gallows, merely an ol' miner chasing Calaverite under the Gallas where a "raise" doesn't mean extra pay.

Just play along and act like you know what they mean. You'll catch right on.

Teasers

Mule Work on the Bull Gang Again
"I found out everything was heavy but the paycheck and the more I did, the better they liked it." ~ Thomas Ronan

Hunting Season's a Good Excuse to Say Goodbye
"We took off for three days, come back and he was waiting." ~ H.V. 'Bud' Peiffer

Communication, Hairnets & White Boots
"It was a very mysterious atmosphere. Dangerous work, and not everybody's going to make it out in one piece." ~ Brock Gay

The Skipper, The Mule Skinner and The Dump Rat
"When I was eleven years old, I started working on the top car on the Vindicator." ~ Michael 'Mickey' Nothaus

You're 'Da Man
"Everybody likes you because their life is in your hands. Those miners would bribe ya t' pull all their muck out, so they can get on with their day." ~ L. R. Myers

Well, It Can't Go That Far
"It was a big truck tire, and that's a very steep dump up there." ~ Jack Lorenz

Mining is a Good Way to Go in the Hole
"This mill, all of my hopes, all of my dreams, all of my efforts just went up in smoke. I just almost felt like giving up. Giving up living." ~ Dale Weaver

Commander Drum, Beaver & Young Men End the Era
"I liked that underground life. The work was hard, but it was neat because not very many people had the foggiest idea what was goin' on down there." ~ Brian Hayes

Beans on the Table
"The man said, "You will," and I said, "Yes, Sir." ~ Edgar T. Hunter

Channel 9 News

"The first thing you need to know underground mining is that your partner and each one of you watches over each other's back. The most important thing."
~ Brian Marshall

Riding the Rim

"When we crawled up that manway, seen all that blood comin' down there, we knew what we was gonna find." ~ Jim McGee, Sr.

The Mad Mucker

"You'll never muck bound us," and the old boy in charge said, 'Oh yeah, we will.' They bet a gallon of whiskey on this." ~ Alfred E. 'Gene' Leaf

Widow Makers & Hand Cars

"Underground is such a strange thing." ~ Henry 'June' Hack

Tour Guide as Timber Man

"I was working up at the Mollie Kathleen as a guide and another fella came up and said, "Would you like to work for us?" ~ W.E. 'Ed' Grosh

Thanks a Lot, Bastard

"I was pretty up front about the fact that I didn't know much. They put me with this old timer. He was just an old drunk, but he was a pretty damn good miner." ~ Gary Horton

Tramp Miner to Mine Owner

"I was probably seven years old. We were rounding up cattle, so I got the flashlight and climbed off the horse and went in there and wandered around."
~ Rich McNeely

The Muckin' Machine's Hittin' the Boss in the Ass

"Mining's pretty much the same. You drill holes, load 'em with dynamite, blow it up, clean it up, put down some track and air line, then do it again."
~ Chip Huffman

How to Buy Your First Car

"If the high school kids couldn't hack the mines they had to eat a lot of crow. I didn't want to eat crow and after ten, twelve days I broke in."
~ Richard 'Dickie' Grater

Every Job But Boss

"I've run machines. I've laid track. I've run loaders. I've done everything there is to be done underground." ~ Willie Roy

Charmed, I'm Sure

"I heard them comin' back, they banged on the airline or something. And when they got there I said, 'Don't pull the chute, I'm in it!" ~ Ted Johnson

High-Grading vs. Pocket Rock

"Every miner high-grades. Some people do it for different reasons. My reason is my status symbol of the mines I've been in." ~ Guido Honeycutt

Hardrock Lover

Under clouded veil of secrecy
In the deepest dark of night
They met in longing desire
Timeless their midnight flight

Calaverite her name
It coursed through his veins
Her heart was made of stone
Silver-gray her eyes glistened by day
By night they made his blood run

With pick in one hand and shovel in the other
He delved down deep in her soul
Further he went overcoming resistance
'Til he found what he'd been looking for

He trembled slightly seeing her there
As he did when they met in this way
Her naked beauty and splendor
Made him shoot two rounds that day

~ Sylece '03

Anchoria Leland Ore House

Mule Work on the Bull Gang Again

"I found out everything was heavy but the paycheck, and the more I did the better they liked it." ~ *Thomas Ronan*

The working conditions, I hate to talk about them. We were slaves, and wages was three dollars a day in them days. At the least little thing, they'd fire ya. They'd fire ya for nothing. Lots of men got killed, but they wouldn't timber the ground. When I went to work we had to have what you called a miner's card. The mine owners, they owned everything. Then in the early days men started the Union here. The miners went against it, and these Union men was run out of town. Now where the old auditorium was, they'd bring the Union men in there and they'd have a tub of water right down, you'd drink like a cow. Then they'd put them in a railroad car and they'd dump them off in a prairie in Kansas. Some of them couldn't come back for seven years.

I found out everything was heavy but the paycheck, and the more I did, the better they liked it. And the least little thing, they'd fire a man. It's so different today. You had to work, work, work to get out so many tons a day. It was real tough. Of course, things were reasonable on three dollars or three and a half. You'd eat better then than you can on these big wages today.

When I went to work in 1918, wages was three dollars and twenty-five cents for eight hours. And then shoveling coal, and lime and mule work, it was three and a half. Well, then after a number of years I got to be a mill operator, wages was four dollars, and it's surprising what you could do with that four dollars. I bought a house on it. As my father died young, I'd inherited the whole family. I sent my sister two years to college on four dollars a day. But today, I don't know, the way things is going up, it takes ten dollars today to do what one dollar did then, in food and clothing.

Did you know any of the rich or important people who visited Cripple Creek or Victor?

Well, I knew some. Mr. Stratton, he was a carpenter, and on July 4th him and his donkey come up on that mountain, and he dug around and found two mines. He called one the Independence and one the George Washington. He found more every place. He sold that mine for eleven million dollars. And the fella that made the sale, Vernon Reed, he got a million dollars commission. He put it in Salt Creek (oil) and made another several million out of it.

Then I did know some lessors. Fella by the name of Harrison, him and his partner, on the W.P.H. made three hundred thousand dollars. And I met him twenty-five years later, and he was plumb broke. Old clothes, trying to make a living. Then there was Cornduff Duncan, another 'dang mine, they made an awful fortune. But they just squandered their money, and Cornduff was a very highly educated man, he come from the nobility of England. He'd lend money to all his friends. Pretty soon he was broke, and he wasn't eating, and he was leasing on the Vindicator for a whole year. No pay, nobody'd pay him back. So he goes on the Vindicator dump, he puts two sticks of powder under his head and blew himself, and killed himself. His will was written in Latin, and he give the list of all those that owed him money, that they could pay three or four cents on the dollar, so they could bury him.

Then there's several others made millions here. But they didn't spend any of it here, most of them, they took it out as it came. Boy, there was a lot of fellas. Nelson Crimedon, he was the manager on the Golden Sampler. When I was a kid I used to cut his grass, he was really nice. Well, there was lots that I didn't know, lots made money here. But you don't hear about those that lost money.

Now on South Fourth there's the Saint Patrick Mine. Executives of the Clark Thread Company put in one million dollars, they sunk the shaft and they didn't get a pound of gold. You can hear about them. I wasn't very well acquainted with the millionaires. I did know lots of poor people.

The men started the Union, they was just gettin' three dollars a day then, for ten hours work. And the mine owners said, "No." Well, pretty soon they started to kill a bunch, and lots was shot. They had the militia in here. Well, when it was all settled, the mine owners won.

Now when I went to work in 1918, before you could go to work you had to go to the Mine Owners Association and get a card. If you was a Union man you wouldn't get it. You couldn't get a job 'til you had that card. As soon as you got laid off or quit, why, you had to turn that card back in. If they caught ya stealing high-grade ore, they took that card away and you couldn't work any place in the District.

Well then, when it was finally settled, the men got three dollars a day, and eight hours. But the mine owners was awful mean to the men. Well, there was men from all over the world here. When you'd go rustling for a job, there'd be fifty, sixty, eighty men right there.

What different kinds of jobs did you have in the mines?

When I first started I was fifteen years old; I was on up at the Vindicator. Well, the law said seventeen, but I was such a big fella for my age, and when I got my card I said seventeen, I put it over on them. I was desperate for work. You had to be seventeen.

I was firing a roaster up there. See, they got the concentrates, and we'd roast them, and we'd load 'em in a boxcar. It was three shifts, we'd work around the clock. Well, that run one year and shut down.

Then I come down to what they called the Independence Mill. I was what you call the Bull Gang Man there. You know, loading lines, and coal, and odd jobs. And then later I got into sampler. I was around the crusher and that kind of work, and then I got to be the Foreman in there. Oh, it was an awful job. Well, then I got knocked out with dust and I went on the repair gang, and then I was a mill operator, repair man, and the like of that.

Well, December 1, 1928, the mill closed. I was there nine years. Then I went up to work on the Portland Mine. I shoveled underground, and timbered, and was what you call the cager, and then I run motor cars. Then they run out of ore, I was laid off again. I went over to the Ajax, and on the Bull Gang again, that's where you started. Then I was fire man there. And then they had to pull ore uphill, I was on that hoist; I guess that was seven years I was there. I loaded eight thousand cars, 300,000 tons of ore.

Well, then I quit there and I come back to the Portland again, odd jobs, I was finally the ore house foreman. In '49 all the mines closed. They pulled up the railroad to build the Carlton Mill. So I went over there, it took two years to build it and in '51 the mill started. Well, January 1st, 1962, the mill shut down. I was a repairman, pipe-fitter, plumber, finally a mill operator. When we got everything cleaned up, the Super says, "I'm sorry, Tom, this is all of it." I said, "I'd of never took this job, it only lasted 13 years." (chuckle) Well, then I was too old for another job, so I didn't have too many jobs, but when I got a job it seems I could hold it.

What was it like in the town of Midway?

The town of Midway was quite a place in it's time. That's where the mule skinners stayed, and there was 700 horses there at one time. They pulled ore wagons, and grocery wagons; there was a lot of fights there. But now all that's left is that old place that used to be a boarding house and then was a saloon, it's pretty well wrecked.

What are "mule skinners?"

Well, that's people that drive horses. See, before they had trucks, people would drive horses. Some would have two horses, and then hauling this ore, there'd be four or six horses. Then the driver, they always called him a mule skinner. They had'a get up early in the morning, and oh, it was hard work. They mucked that ore into a wagon and then they'd come down and have to muck it into a boxcar that was high-grade, so nobody'd steal it, and then muck it back. They had to make so many trips a day, and lots of them would wear out a shovel in six weeks from all that shoveling. It took a mighty husky man to do that kind of work.

In those days you had to get out and work at an early age?

I'll say. I was born in this camp. I've been making my way since I was eight years old and when my father died I didn't inherit any money, I inherited the family and at fifteen I was working at the mine and I stayed with it. My mother lived to be 95, and that's the way life went. But all us kids worked, we had to earn our own spending money in them days. And it was hard and heavy.

There was a place called Bull Hill, I was never out that way very much. But there was one town, right above Independence, Altman, that was the highest incorporated town in the World. Two thousand people lived there. That's all gone.

That's where the major part of the miners were?

Yes, up there. That was a wicked place. The miners were ahead of their time. Their forefathers wanted a little better working condition, a little more money, and them millionaires said "No, no, no." But there was so many thousands looking for work, all over the world, no place to go, they really took advantage. In one year there was thirty millionaires made in this town. But they didn't leave any of the money here.

Grace Greenwood Mine

Hunting Season's a Good Excuse to Say Goodbye

"We took off for three days, come back and he was waiting."
~ H.V. 'Bud' Peiffer

My dad, Lou Peiffer, was a miner. He graduated from high school in 1924, here at Cripple Creek High School. The summer after he graduated he went to work for some people by the name of Wilson. Dick Wilson also graduated that year, and they started mining up on Gold Hill. The name of the mine was the Grace Greenwood. And so, one thing led to another; he just kept working there. It was a funny story because one of his classmates that graduated went on around the country somewhere and came back, 30 years later, and my dad was still on the same mine. That's the mine he leased and worked on all his life. This friend of his said, "Well, you've moved a little ways." They'd moved the ore house and the hoist room about 40 feet over the years; they'd set it back. He said "Well, yeah, you've moved a little bit, about 40 feet."

That mine goes down to the Index, which is in Anaconda Gulch. When you get to Anaconda, where the train ends, that's Anaconda. That used to be the old town. Straight north of that, the Anaconda Tunnel went back in under that whole mountain, Gold Hill, it's called. The bottom of the gulch is called the Index Mine, and the top where they have a headframe was the Anchoria Leland. Then there was the E. Porter Gold King, and the Grace

Greenwood. Those four mines. So Dad, that's basically where he was, from 1924.

When they shut the mines down during the war, he didn't leave. He sold out the lease, finally, in '48 or '49, and then he moved to Uravan, and went to work in the Uranium mines. He was over there from about '49. I went on there, too, when I was first married, and worked in the uranium mines, in Nucla and Naturita, Colorado. Uravan is where they've got that big Superfund cleanup of all the so-called cancer causing agents, and a lot of the people from Cripple Creek and Victor went over there, because it was booming.

It was a whole different type of mining. In fact, the first vertical shaft in that whole western plateau was sunk by my dad, and a fellow by the name of Julian Davis. It was a vertical shaft, which all of these over here are vertical shafts, but over there they were all what they call inclines. They just go down at about thirty degrees, and you were never more than a couple of hundred feet underground over there. Didn't have any deep mines like here.

Then I worked on the Ajax when I came back. I worked the Deadwood, worked for Lloyd Stoner and Jerry Musik on the Deadwood, contracting. A fella by the name 'a Pete Peterson, and I. He went on to be a Denver Policeman, and I went on to be in the Border Patrol in 1956. We left here and went to El Paso, Texas.

I worked in the mines about, off and on, six years, seven years, something like that. Contracted, and worked for days' pay. I remember a machine man, running a stoper or a jack-leg, got a dollar and a quarter an hour. That was pretty good money back in those days. Just a laborer got ninety cents an hour. (laughter) I remember the money.

Then I got in on a contract driving a lateral on the 3100 level of the Ajax, with fellas by the name of Mickey Fleetwood and Jake Bauer. The three of us had a contract, and I was making more money than the Superintendent of the mine. But, of course, we worked our way out of a job, because we finished it. That was quite an experience. I think it was in '54.

Then I worked for a period of time as an assayer at the mill, when the mill was right there on the road between Cripple Creek and Victor. The Carlton Mill. I worked summers helping building that mill over there when I was a Freshman and Sophomore in high school. The mill, that was a good move, but then the mines closed in '61.

Why did the mines close?

Oh, just the price of gold was fixed back then, see, and you couldn't make it. The money, where the good ore was, and that's all you could ship was the real good stuff, was like in the Ajax and the Cresson, and that was all getting down so deep, it cost so much to mine it. It was prohibitive to mine it, so they just made the decision to close it down. And then, I guess, a lot of people left here during the '60s, after the mines closed down.

I was away in the Border Patrol; we were living in Spokane, Washington, when the mines shut down here. My dad had been working running the hoist over on the Ajax, when they shut down. So, he was affected. Then he went to work for the County, he took care of the boilers at the Courthouse. So that was good, he got out of the mines. But he died of lung cancer, probably caused by those five, six years over in Uravan, uranium mining.

What's it like to mine uranium?

Oh, it's soft. Those mines, you can smell the good ore. That uranium was hot. You'd go down, you'd come out from underground, they'd put a Geiger counter on you, and it would light up that Geiger counter.

Did you eat lunch underground there?

No, we always come up for lunch. There's too many little bugs down, that you'd eat. I don't know what they are, but they're little flying bugs that are underground, even over here. They come in on the lumber that goes down, where they'd have the ties and the sprags and stulls. Everything. Then they'd just multiply. There's a lot of little bugs underground. Yeah. Well, we used to call them shit hawks. (laughter) We always had a designated drift where everybody went to do their business, that's where they mainly were.

We loved to come up to eat lunch. Over there in uranium, it was only a couple minutes up to the top. You know, it wasn't a big production. When you worked underground at the Ajax, or the Cresson, you ate lunch underground. You went out to the Station, and ate lunch. Because there's always lights out at the Station. It gave you a chance, they always had battery chargers to recharge your lamps and things while you eat lunch. You took your lamp off, plug it into the charger, and then you could go the rest of the shift.

Time went fast underground. It really did. I didn't mind the work at all. Seemed like you'd just get down there, and it's lunchtime. Then we always went up at 3:30. It was 7:00 to 3:30, always. Most of the mine's were on that schedule.

They just ran one shift?

All they did at nights on the Ajax, and sometimes I would stay over and get some overtime helping, was lower timber and things that they needed for the next day.

Lower it how?

On the skip. You'd just load the timbers on the skip, and somebody'd be down there to unload it. You'd just keep getting all the timber that you could down each night, for three or four hours, and that would take care of the guys for the next day. And rail, if you needed rail, it all came down at a different time. Because during the day, the skip was going up and down with ore, see. So you had to wait 'til night, after everybody was out, to lower your supplies for the next day. Including dynamite, caps, everything like that. You'd put everything together above ground and then lower it for the guys for the next day's shift, on each level.

As I remember on the Ajax, they had a stope on 2400 that was real active, and 26, 28. A lot of activity down on 31'. Later, after I left mining, they sunk a big winze on 3100 at the Ajax. A winze is where you just go off the level, you're down 3100 feet and so then you just sink an upside-down stope that just goes straight down for a couple hundred feet. That was a winze. It was all on that Bobtail vein, that is actually under the Victor Cemetery. If you went straight up from that Bobtail vein to the top, you'd come out right under the Victor Cemetery. It's never been mined that much.

But they were fighting water on the Ajax all the time. In a wet year you had to pump water all the time. The Carlton Tunnel was the 3100 of the Ajax, see, and that drained all mines, supposedly. But, there was water always running a foot deep on 3100. You always wore rubber boots and rubber clothing because it was so wet, and hot down there, too.

Did that effect your skin?

No, some people got rashes and things, but I think a lot of people didn't shower that often. You know, they were still living in the 'Saturday Night Bath Era.' Once a week, whether you needed it, or not. But, we had a big shower room at the Ajax and everything. It was available. I mean, you didn't need to not do it. A lot of people, it was just automatic. You took a shower when you got outta there.

We had a basket on a rope, and you put all your clothes in your basket and tied it; get it up off the ground. Because they would come in there and clean during the day, so your basket would be up above. It wouldn't bother anything if they went in and washed everything down. There were some lockers on the other side, I remember, too. I always just used a basket. But that was quite a mine. They had 60 or 70 people.

Then, the Mollie Kathleen was a mine that my Uncle Hereford and my dad both worked. They worked at night up there, because it was a tourist mine during the day. They went up there on that Black Diamond vein. That's when the mill was shut down, and they had whatever ore they would take, they would truck it over to Leadville, to the smelter over there.

They closed the smelter in Colorado Springs when they took out the Cripple Creek train?

Yeah, they tore the train down and then it was two, at least two, maybe two and a half years, before they had ever started doing any smelting at the mill over here. They shut the smelter in Colorado Springs down in '48, started tearing out the railroad in '49, and started building the mill in '50. I don't think they opened the mill until '52. So, that was the time frame of the mill.

There was a lot of people trying to promote mines around here. That's what they were: promoters. They'd just go to Denver and try to get people to invest in the mines, and you know, the stock market. United Gold Mines and Golden Cycle owned 95% of the good mines around here.

The Mollie Kathleen and the Gold King would be an exception. The Gold King, where gold was discovered, that's right there below the Mollie, and it's a 1000 feet deep, too. It connects with the Mollie Kathleen at the 600-foot level. They built what they call a bulkhead, to keep the air pressure; to keep the gas out. The Mollie and the Gold King were both very gassy, and that gas was just silent. That's why you always had a carbide light burning. If that light went out, the oxygen was going out, too, and you had to get out of there. We always hung a carbide light wherever we were working, to where we could see it at all times. Or candles. Even candles.

Later on they had ventilation systems, but for years they didn't have anything. Later on, like when we were doing that big drift in the lateral over on the Ajax, we had to have air vents come in, run by a compressor. But that was just from being dead air, it wasn't gas.

Back in those dead-end drifts you'd shoot twice a day. You'd put on dynamite twice a day, and that's what they'd call a hot muck pile. A hot muck pile, you could just smell that dynamite, and it'd give you a terrible headache. So, you'd have ventilation circulating in the heading where you're driving. The rock, the wall is right here, and that's where you drill. That's where it was just dead air. You could get terrible headaches from that dynamite.

A lot of people, if they didn't like somebody, when they was taking lunch or a little nap or whatever, they'd slip a little nitroglycerin, which is dynamite, just unwrap it, take a little bit of that and put it in the sweatband in his hat. Heh. You don't last long with that, boy. It gives you a terrible headache. You'd finish your shift, or you'd have to go out and give the bells that you wanted to go up. Then you'd have to go and punch out and that cost you, because everybody was on a time-clock for the mine. Unless you were a contractor.

If you were a contractor and you did all you wanted to do that day, you could go up at two o'clock, or one o'clock, or whatever; whenever you wanted to. Contractors always got first preference as far as the skip goes, for men and stuff. And their muck. All their ore, or non-ore, or anything that was hoisted always went up before the company's did. They had the first ride out of the station to get your ore, or muck removed.

If you're contracting, what you're trying to do is get two full rounds in and out a day. But, what happened normally was it was like every other day, you'd shoot twice, muck out once, and shoot twice. So, you're actually getting a round and a half in every day. But some days, if everything went right, you could do two. It was six-foot steel, so you were making twelve foot a day. That was excellent. On that contract we got $15 a foot, and that's good money, split three ways. The three of us had the contract.

Did you have to muck it yourselves?

Oh, we had mucking machines. Just one, Eimco mucking machine, a 12B, I think it was, might'a been a 24. We had three jack-legs in the heading. The guy that was running the machine in the center, he did the cut, what they called a cut. The two guys on the outside did the rib, and it was like there'd be a five hole burn cut and the relievers. He'd do about nine to eleven holes in the middle, and we'd do about eight on the sides. If things went well, you could drill out in an hour. Then you'd shoot it and go back in. That was mining.

Now the guys up in the stope, see, they took themselves up right on their own muck. You'd set up what you'd call stopers, drills on a series of lines, called 'em sprags and stulls. You'd go up there and set your little platform up. First you'd have to chisel into the rock to make your logs fit down to where they're solid, to where you'd climb up on 'em. Then you'd get up with your stoper and drill vertically, straight up. Then when you shot that round, the muck would all come down. You'd go up and pull the muck down from the level; it was called shrinking. You'd shrink your stope down, and then you could get back up in there, and do it again. The muck all comes out and then down on the level; you'd take out enough that you could still go up.

Some of those stopes were like two and three hundred feet high. That one on 26 of the Ajax was a real hot stope, that was good ore. It went to 28, almost to 29, and you could see those guys, those were all company guys way up in those stopes. It's a long ways up in there. They had people down, taking the muck away. Tramming it out to the station. I think they sorted most of that ore. They had an ore house there on the Ajax and the skip would come up and dump down into the ore house. That would be on a belt type of thing, and somebody'd be throwing out the solid granite and keeping the good ore, and separating it. It was trucked away to the mill. That was quite a process over there on the Ajax.

And the Cresson, that was a good mine for years. I worked there two weeks one time, filled in for a guy. Franklin Ferguson, and a guy by the name of Bill had the lease over there, and I can't remember who it was that had to take care of his family's business, or somebody died or something. They said, "You wanna come over for a couple 'a weeks?" So, I worked there at the Cresson two weeks, and that's all the time I was ever over there. That was in the stope. But they were leasers. That was the difference between the company people and the leasers, and they had pretty good ore on the Cresson.

Did some mines lease out certain levels, and then have Company men on other levels?

Yeah. It depended. The Cresson was more like that. The Ajax was more a company mine, except for the contractors, which developed the mine by taking a lateral out and doing that. But the Cresson, as I recall, they had a lot of leasers over there. Bert made some real good money after he got back from the Navy. Like '45, '46. Him and Buck had a lease. Then there was Hanson, he had a lease. There was just quite a few leasers, and those guys made big money right after the war. Most of those guys all ended up in Uravan a few years later. Hanson, he just retired; he had enough money.

When I was a little kid, I must'a been in the 7th grade, about '46 or '47, and Jack Schwab had just opened the Cottage Inn. We got out of school one day, and we went downtown and we saw what was happening. And here's the Bunkers, and the Bielz, I can't remember all their names, been drinking in the Cottage Inn all day. So, they decided to see who could throw a silver dollar the furthest across the street. There was a vacant lot, and then there was the Iland Department Store. Us kids were over there to pick up those silver dollars. I mean they did it for like an hour, just seeing if they could get the windows in the building across the street because they were all vacant anyway. We thought we'd died and gone to heaven, we were over there picking up those silver dollars.

The Depression never hurt Cripple Creek. The mines were going in the '30s, and if you wanted to work, there was work. And decent pay. In fact, up until they shut the mines down the first time, was in 1941, because of the war, actually the order didn't come until July of '42, before they actually shut the mines down. Because they wanted all the miners from here to go to the copper mines in New Mexico, Arizona, anywhere they could mine lead, zinc. Leadville was going real big in copper and molybdenum. But, there was no need for gold, so the Federal Government shut the mines down.

Did people mine here, anyway?

The mills didn't accept your ore. See, every mine, even if it was independently owned, was supposed to be shut down. I think there was a few little operations going of people who never left here, and they'd just stockpile. You had to have a place to ship it, to get your money. But I think they continued, and that Order didn't last. I think it went out like eighteen months later, I don't think it lasted for all that long. But there for a while, boy, just everybody was up and moving and had to go. Then they all came back. It was another good ten years of good mining, probably from '42 to '53. Right after the war, '45, '46, '47, that's when it was really going good around here. Then it kinda started to tail down, and along comes the uranium, and everybody went over there.

Miners are pretty nomadic, see. They like to follow... my dad went to Leadville before he ever went to the uranium mines. He spent a winter in Leadville, 1948. At the Resurrection in Leadville. Then he worked on the Yak Tunnel for a while over there. That was tough country. My brother and I were both still in school here. I was in the 8th grade, and Jack was a Senior. Neither one of us wanted to move to Leadville, so I moved in with my Grandmother here, and Jack, he just had a few months to go 'til he graduated high school.

What did you think the first time you went underground?

I liked it. The first time I ever went underground was right there on Dad's mine, the Grace Greenwood. When you're a little kid, they'd put you in the bucket. As you got older, then you'd ride the rim of the bucket. Hold on to the cable. There'd be two of you, one on each side of the bale to balance the bucket when you went down. I remember going down. The Greenwood had good ore on 200. I think it never got below 400 feet, but it went out from the shaft several different directions. They mined for a long time. But that's the first mine I was ever underground in. I must'a been, oh, just about six, eight years old. Something like that.

Did it seem natural to you to go underground?

Oh, yeah. When you grow up, you know, even as a teenager, I worked up there, too. For my dad. You didn't know anything different, and you weren't smart enough to be scared of anything, so there wasn't anything to be afraid of, really, unless you made it that way. Unless you were careless.

What kinds of things did your dad have you do?

Oh, I always run a machine. In fact, I had been on a machine up on the Deadwood when they first came out with the jack-legs. I talked my dad into a jack-leg, because they were so much easier than any of those other things. In the first place, there was no Leyner. They called them Leyners; you had to set up a big 'ol column that was adjustable, it was hydraulic. You had to set that up, and then you had to get this great big, heavy machine out on the arm of this column to drill. And you had to move that about four times while you were trying to drill a round. With the jack-leg you'd just walk in and start drilling. You'd walk here, over there, down, however. Dad didn't like it at first, but he liked it after we taught him how to run it.

Were there accidents in the mines where you worked?

I was involved in one fatality rescue. A fella by the name of Cub Peterson. Big 'ol boy, that's why they called him Cub. He looked like a cub bear. He was working on this tunnel right below town. I can't remember the name of it, it's in Anaconda Gulch. Anyway, he didn't show up at the pool hall, he always came to the pool hall after he went home, you know, five, six o'clock in the evening. He'd come down to the pool hall. Butler's Pool Hall, right there where Bronco Billy's Casino is now, and somebody says, "Well, where's ol' Cub?" Nobody knew. So some of us young guys were shooting pool, so we said, "Well, we better go down." Because somebody knew where he was mining and he was working by himself.

We went down and got some lights, and went back in and you could see where he had a tramcar and was going over an old stope, or a big winze. He was up on top of it, and he was taking the tramcar over it, and you could see where the rail gave way and the logs all caved in, and he was down at the bottom. About 150 feet down. His light was still burning, so we knew that's where he was.

They got the Sheriff, and some other guys, and it was like nine o'clock at night and Doc Denman, the local doctor was down there. He was also the coroner. So they all got their heads together, and decided there was no reason to go make a big effort to rescue him that night, because 99 times out of a hundred you'd be dead. And that light hadn't moved or anything for hours. So, the next morning they got the State Mine Inspector, everybody, all the figures, and somebody took a gurney and went down and got him out. The next morning. They were right; there was no reason to do it that night. We were the first ones down, and saw his light and stuff. Saw what happened. You could see what happened. Then we went and got everybody.

Then a good friend of mine was killed on the Ajax. This was later. This was in the early '70s. I can't remember the name of the company, but it was doing some mining on the Ajax. Paul Jones and a couple other guys, they got gassed. They got on the skip, and they hit the bad air, and it killed 'em. They found 'em right in the skip. That's the last fatality around here that I know of in the mines. That was just because they didn't think there was any bad air. And usually out at the shaft there isn't, because you've got drafts and things. This was at night, they were working nights. I don't know what caused it. Something caused it, and they were unaware of the gas.

The hoistman didn't know. It wasn't his fault. Any good hoistman will tell you, they'll sit there for a week until they hear from those bells to move that skip. See, just like if you take that mine tour up here at the Mollie Kathleen. You got all those tourists down there, and that hoistman's up on top. He can see by the cable that you're loading people on the skip, because the cable moves a little bit up on top, and they're a thousand feet down there. If that guide doesn't ring the bells to go up, he'll sit there forever. And he has to. Or, to go down is the same way. When you load on that skip to go down, the last thing the hoistman does, is he hears. He can see the guy right out there giving the bells to go down, but until he hears them in the engine room, why, he doesn't do it. That's a very good rule.

The only time I ever experienced anything like that, we were dumping big two-ton cars into the skip on the Ajax, and it got away from us and went right in the skip. The whole two-ton car. So, we went and got help, and that skip was still sitting there with that big ol' two tons plus of rock, in the skip. Because the hoistman didn't move it. The truck of the ore car was sitting up over the edge of the skip, so if he'd 'a moved up, he would'a tore out all those timbers. We got back with crow-bars and more people, and jacks, and finally got it out, and of course dumped it and rang him up. That was an hour later. (laughter)

They could have had better communications, like telephones or walkie-talkies. They had them during the war, but the miners never got in to that too much. In the first place, all the telephones and the electric lights and everything, it was so wet underground, you lost a lot of them. You couldn't hear a lot of things. And so, you just depended upon your signals. That's been since the early days of mining, and it hasn't changed at all.

I've ridden with some hoistmen that were a little more jockey-types, and others that are real, real slow and real, real cautious. Then other guys that would let 'er go, you know. Talk about Gene on the Cresson. That was the fastest hoist, fastest skip, in the District. Well, everybody was out all night, usually, with a hangover in the morning. You'd get on and he'd let that first skip go down there (laughter), get their attention, sober 'em up.

That hoist was fast because of equipment. It was a fast hoist. I'll tell ya, the hoist on the Ajax, there was a guy by the name of Otto Sours, run the hoist. I never will forget what he looked like. An old guy, he could'a had a heart attack at anytime, he was up in his seventies. And he was running the hoist. But he'd been running it for thirty years.

As close as I ever got to being hurt in a mine was right there at the Ajax, we were working on top. It was a huge hoist, 3000 foot of cable, and it was as big as this room. What they'd do is periodically take and let all the cable off, take the skip off and just let the cable roll down in the shaft, real easy, let it go down. Then you'd take gasoline, and you'd scrub that drum of the hoist. It's like a ten foot diameter drum, and there's a pit underneath. You'd take this white gas, and that's called cleaning the hoist drum.

I was up on top with another guy, with a come-along, and we were lifting one part of the hoist up. It was servicing the whole hoist. This kid over by the door was smoking a cigarette. We had been taking the white gas out the door and dumping it, and of course had been spilling a little bit. He flicked his cigarette and white gas just came flashing into the pit, under the hoist, and just exploded. And we're up above, so I jumped off.

'Course, I was young, jumped off about 22 feet from the top up there, down to the cement floor. The one kid that was in the pit, one of the Lorenz boys, Gene, he burnt his arms really bad. So, we took him to the doctor over here. But that could have been almost a disaster for several people. We went back and finished up cleaning up the hoist, and got it back up. Then you gotta rewind that cable, it comes back up and that's when you do it real slow, and you got a guide thing and you just guide that cable back on where it wraps evenly. That was a huge hoist. Well, anything that holds 3000 feet of cable, probably more than 3000 feet, probably more like 4000 feet, has to be big. But that was on top, was the only time I ever come close to getting hurt.

Did people smoke underground?

Well, I didn't smoke, anyway. Most of them chewed underground. Copenhagen, or tobacco. Some of them smoked, you could smoke underground. It wasn't that bad if the air was good.

What jobs did you do underground?

Well, run the machines. Mucking machine, jack-leg, did the powder, the caps and the powder. Wired the rounds to where the shots go off the way you want them to.

The shorter fuses go first. You want, when your heading is like 6x8', or whatever it is, and you drilled your holes, you've got a concentration of holes. What they called the cut. It was either five or seven holes, right in the middle. All these holes, and you shoot that first. That all comes out, and then the rest of it caves into the middle, because the powder breaks toward the path of least resistance. So, that's how they'd shoot these rounds. You'd cut 'em right off. When the cut goes off, it goes off all at once. You wire it to where it goes off just as a single shot. And then, the rest of them are timed to go off.

It all shoots and you'd have your muck pile, six by eight and oh, three-quarters high as that window, probably. About six foot, seven foot out away from the heading, and then you'd go in and muck that up on what they called temporary rails. You put your mucking machine in on these larger rails. You'd take these rails, temporary rails, and take a double-jack and pound them into the muck pile, and then your mucking machine had something to be on, the rails. Over in Uravan, they had rubber tire machines. But not here.

Was it all rails?

Oh yeah, quite a little railroad system. We had motormen; we had a big motor that was powered by nothing but batteries. It was electric, with a switch just like a motor car, like a street car. Then of course they had some powered by air, locomotives. That was made right here in town by Fry & Company. Couldn't have gas engines, too much exhaust. Diesel they did have, but that was terrible underground. That's why they went to electric, or air was the best, of course. But then you had to stop and air up your locomotive.

What do you think of Tommyknockers?

I don't think anything of them. I know one thing; you can hear people in other stopes that might be a couple of hundred feet away. The sound carries. If they're trying to bar down, what they called bar down, you'd go in to a stope and if there's loose rocks you take a bar and knock those loose rocks down. You can hear 'em do that. But as far as Tommyknockers, that's some dream from Ireland or something.

Did you have much to bar down in the hard rock?

A lot of time you didn't have to bar down, it'd be just slick up there. The slabs that have hurt people up here have come off the hanging wall, or the sidewalls. A big slab of rock, once in awhile, would just pop out from the pressures of the earth. And, if you happened to be in the wrong place at the wrong time. Marvin Cook, a good friend of mine, got hurt on the Cresson that way. He was just up in the stope and the whole wall caved in on him. Broke his arm up really bad, it really hurt him. But that was just from pressure.

I have seen, you can take a load of ore cars out behind the locomotive, go out to the station, dump, start back in and see rock all over the track. It's all popped out from the sidewalls. Or, not very often, from the roof, you might have a rock that fell from the roof. But these big slabs, as big as that chair, just pop right out in the middle. You'd have to get off and take a bar and bar 'em, break 'em up, get 'em off the side before you could keep going.

It never scared you?

No, it was just a freaky thing if you ever got caught with one of those. A matter of timing. In the wrong place, at the wrong time. Marvin got hurt, and I've heard of several people hurt when the rock popped. This is good hard rock. Other places you'd have to go in, what they did was drill holes up in, all the way around the heading, and Gunnite those holes to keep the roof from falling in. Around here, you'd never have to do that. So, it was a lot more expensive to do it, like in mines in Leadville. Like the Climax. I've heard everything had to be Gunnited over there.

Gunnite is a cement, is actually what it is. They'd just drill holes and cement to keep the rock from caving in on them. Then they'd use nets and they'd do a lot of things. There are places, right here in the District mines, that they call unstable rock. There'd be a big dyke or something come through where the ore is, and that would be bad rock. It could be caving rock. But you knew where it was, and you'd usually just take timbers and timber it off. Because you'd have such a problem with it all the time, with rocks coming down, that you'd just build like a snow shed, like you do on the highway. Just build that to keep it from coming down on the track.

You had to do that because those things, you just spent all your time cleaning up. A couple 'a days of timbering and that rail through there would be fine.

Did everything have to be timbered?

No, no, just where it was bad. I mean, like the Ajax probably had, on 26 I think they had a place where they had bad rock. They built things over the track. And the Cresson did. The Cresson had a little worse ground. It was different than the solid granite of the Ajax. In the Ajax, everything was solid granite except for the ore. It was gray, and the streak, you could just see the streak. It could be a foot wide, or three inches wide, or down to an inch. You could tell the ore immediately. Otherwise, it was just Rosebud Granite. When you're drilling it, if your hole was running red, the water coming out running red, you were just in hard granite. Then if you hit a softer spot, the gray, then you're in to where the ore is. That was the easy way to tell.

Was there a lot of high-grading?

Oh, some people took some good specimens, but it wasn't like it used to be in the old days. I guess the last real problem that they had was when they hit that vug on the Cresson. That was under twenty-four hour security, because it was so rich. But on the Ajax, yeah, you'd have good ore. And then, like my dad's mine up there on Gold Hill, the Grace Greenwood, that was a lot of free gold. They called it Peacock gold, or free gold. You could actually pan that gold up there and see what it was. What they called Breccia. Breccia rock; it was soft. There was no granite on Gold Hill. They haven't even started to do Gold Hill. They'll ship that. They'll leach that whole hill. That'll be the best hill they've ever been on.

I remember two old brothers. I used to hunt rabbits up there all the time, by my dad's mine. The Owens brothers lived right across from the old high school. They made a livin' just digging trenches, up on top, north of the Grace Greenwood mine, and mining and sorting the float rock. Just down to bedrock, just the float rock. Move all the top soil, get all those rocks. Look at them, sort them, sack them. Had an old pick-up. And, they made a living doing that. They'd go up there every day and mine those. They had their hammers, and their picks.

I'd go up after they'd get done working there, and I'd go into all those little trenches and hunt the rabbits and stuff. That was a good spot to do it. You could see where they'd break up those rocks and find, it was free gold. There was a lot of free gold. That's the only place. They say Mineral Hill they found free gold, but I never saw any from there. But Gold Hill had a lot of free gold. At the E. Porter Gold King, that was a hot little mine, too. And the Anchoria Leland, that went down a long way. It was years. See, you look at those four mines up there and there's hardly any waste dump, because they shipped everything. It was all low grade ore. Some of it was high-grade. Dad worked there 40 years and he never had any, there wasn't a waste dump. They shipped everything. Everything that came out of the ground.

Do you remember the Anchoria running?

No, the E. Porter Gold King did. And the Index, I don't think it did run while I was a kid. The Anchoria, I can't remember anybody being on the Anchoria; could have been, like say, '39 or '40. Along in there, but I was just too young to know it then. But the E. Porter Gold King, we had to drive right through there to get to my dad's mine. They were only 400 yards apart. The Anchoria is first, then the E. Porter Gold King, and then the Grace Greenwood. So, those three claims were all right close together. That was a great little mine. He made a living there for years and years, and he worked for himself.

You had to buy your air, what they called 'Air,' from Teller County Air. There was a series of pipes all over the District, that carried air that was made by big compressors over by the, I think it was by the Portland or Independence or the Vindicator. One of them over there. All these big compressors.

They sold air to all the leasers and the miners that leased up through United Gold Mines. It was called a split-check lease. You furnished your own materials, your own drills, dynamite, and it was a 60/40 split. You got sixty percent, and UGM got 40 percent. But you had to buy your air from Teller County Air, to run all your machines. That's better than having to buy a big compressor and do it all on your own. There was a guy by the name of Johnny Phipps who was the man in charge of Teller County Air, and he rode a big white horse all over, up to Cameron, Gold Hill, Stratton, anywhere there was any mining going on that used Teller County Air. He rode that airline to make sure nobody'd shot a hole in it, or it was losing air. I can't remember what they paid for air. It was a monthly charge.

When did it go out?

'48, '49, somewhere around in there. Can't remember; maybe it was in the '50s. It was something that all the leasers who were on split-check, like Dad was, didn't have to spend several thousand dollars for a compressor. It was pressurized, and of course you had just a connector, a rubber hose that was to your outlet and you'd just turn it on, and there was instant air. But, it'd only run between eight and four. So, if you wanted to go to work at seven o'clock in the morning, you could go, but you weren't going to have any air until eight. Because they started it up every morning, like at seven, over where the compressors are, and it'd take an hour for it to build up.

Then they'd say oh, we're shutting her on down, low air pressure, couldn't get enough air to drill. That's when Johnny Phipps would have to go find out the reason that we were losing air. Because, you know, there wouldn't be that many people using it all at once, because they knew exactly how much they needed to carry for all the people. But that was quite an interesting thing, I thought.

Were pranks a popular distraction underground?

Yeah, a lot of pranksters. Like I told you about, putting dynamite in the headband. That's nasty. But then the other ones were nailing people's pie cans down, or else welding it shut. Over on the job at the mill that was done all the time. Your lunch bucket was welded shut by somebody'd come by with a bead and tack it shut. You couldn't get into it 'til you took a chisel.

I saw one thing happen on the mill. When we were building the mill we were installing purlins right under the roof. It's like 80 foot to the floor, and on the ball mill level they were pouring the concrete, putting in the ball mills, and we were working above them. There was a crew of about six of us, and I looked around, and another one of the Lorenz boys, Kenny, he was gone. He fell, and he missed those ball mills and that concrete. He fell right in between them, in the dirt. He went home, and come back to work that day. Fell about 80 feet, and just shook him up pretty bad. He'd bit his lip, but he came back to work that afternoon.

His dad, they had so many kids, George Lorenz was the dad. He was a great big guy, great big. He was a real prankster. He was the head superintendent of the machine shop and he kept all the equipment and machine shop, blacksmith shop, everything on the Ajax, he was in charge of all of the on-top operations. He could rewind motors; he could fix anything. He could build anything. Well, he was an actual genius is what it was; he never had any electrical engineering schooling or anything.

One time Warren Hobert, who once lived in this house right here, next door, he worked on top over there. And it's snowing and the wind was blowing and everything, and old George said to Warren, he says, "Warren, you gotta get up there and grease that sheave on the headframe." It's like 70 feet up there to the top. Ole Warren gets his grease can, and his brush and everything, and starts climbing up there. George Lorenz says, "He doesn't know that's a sealed unit." He got up there to grease it, and it's sealed. He come down and he was cussin' old George. He got a lot of people on that when they first went to work there. He'd send them up to grease the sheave wheel.

They had a good mine up there. Monty Grice was the superintendent, the underground superintendent. Like I say, Otto Sours was hoistman. He lived in Goldfield. He had a couple boys. I worked with one of them, Fordy Sours. Well, most the kids around here, at one time or another when the mines were going, they had some exposure. Even if you did go to college, you could come back and make some pretty good money in the summertime, working in the mines, you know. A lot of them did that. But when we left here in '56, there was pretty much nothing going on.

What did people do?

Well, there was still a little mining, but not much. Most of them worked for the County. My Uncle Hereford was a Deputy; he was made County Commissioner. He was Superintendent at the Cameron Mine in the late '30s and early '40s. Everybody thought that out at Cameron is where all the real high-grade was coming together. A lot of geologists thought that down in that area would be where the real mother lode would be found. But I think the Mother Lode was found on the Cresson, in that vug up there. I think that was it. I've never seen anything compared to that. A lot of people had theories about Mineral Hill out here, north of town. There was a few little mines out there, but nothing that ever amounted to much.

Was there much going on in Cameron?

Cameron was a going little town at one time. They had the School Section, and The Cameron. They had several pretty good mines out there. The Cameron was right there where they're doing all that mixing. They've got that machinery in and they're making concrete, pre-mixing rock. Then there was Pinnacle Park, which was a big day-trip on the train from Colorado Springs. You can still see those bear cages and things. Then there was the Forrest Queen Mine, W.P.H., up in the trees back there that you can't hardly see. The W.P.H. was a pretty good mine, and so was the Forrest Queen. But that was oh, half-mile west of Cameron, in those trees toward Stratton. That was one of the last mines, the Forrest Queen. Francis Gunn was the Superintendent of that, right in that saddle between Bull Hill and the American Eagle up there.

Above Independence?

That was the Deadwood. It was a steel headframe. I remember they got hit by lightning quite a bit, sticking out. So they had to ground it pretty good.

What would happen if it was hit by lightning?

Oh, it'd go down the cable and right underground. Fire would be coming down that cable. People have seen that; I never did. Most people wore rubber gloves all the time, because there is a lot of electricity underground, anyway. Not only that, it's wet, and rubber gloves work better than any other glove underground. You had to have boots up to about your knee, and then slickers, and regular jumpsuit rubbers, and then rubber gloves to work on 31 of the Ajax, because it was so wet. There was water running a lot of places there. Some places in the mine was worse than others.

The Deadwood was dry, and when they put that steel headframe up, it attracted lightning in the summertime. Of course, they did immediately ground it real well. And it still got hit. The old hoistman who used to work up there, he was the last guy to live in Midway. He had a little cabin up there. Remember where the Midway Saloon was? Well, he was right across the road, back in the trees. Him and his wife lived in a little shack there.

He was the hoistman of the Deadwood, and the Caretaker, because that road was closed to all those mines that went back in there at Midway. He opened the gate, and closed the gates at night, and locked them. Because you couldn't get in the Deadwood from down in Independence, there weren't any roads up that way. Bull Hill, they had the Hull City, South Burns, North Burns, the Deadwood.

The Deadwood was one of the last independent mines. Jerry Musik and Lloyd Stoner were the two. Jerry backed Lloyd, and Lloyd was hands-on, he worked every day up there. He was the lessee and he hired, oh, I guess at any one time he'd have five or six guys working for him. He shipped the ore right to this mill, out over here.

Everybody had high hopes when they built the Carlton Mill. They really thought that would bring it back. Then the price of gold was fixed. About that time MSHA and OSHA, and all the Federal Government Restrictions on mining started to come into play. They wrote all this legislation that you couldn't do this, you couldn't do that, this had to be this way. The Federal Government got involved in mining in a big way because of the coal mines, and so many accidents, deaths, in coal mines. They just figured every mine, it was a mine. So, consequently, all those rules and regulations were right here in Cripple Creek. It just added to the cost of doing business. Then the deeper the mines got, the more cost prohibitive the mining. Then the wages had to keep going up, and the price of gold was still the same. So, it was just economics finally done them in.

What kind of assay did you do at the Carlton Mill?

It was regular fire assay. We made our own cupels. We had big ovens to melt it, and we did our own samples. Grind it up real fine, take a certain amount. You'd get your samples in and you didn't know where they came from, it was just lot numbers. You didn't know if it was from the Cresson, or the Ajax, or some Lessee. You had to do your assay independently. I worked there for about six months, and then Charlie Carlton called and asked if I wanted to go in on a contract up on the Ajax running those laterals, and I said yeah. So, I made some good money before I got my orders to go to the Border Patrol, and that's when we left, about '56.

What were the working conditions like?

'Course, we didn't have anything to compare it to, you know. You were just lucky, or thought you were lucky, to have a job, and if it was a dry mine it was good, if it was wet, didn't make any difference. You were glad to go to work and people were pretty happy, I think, actually. Nobody ever approached me about getting Unionized. You know what Golden Cycle would have done, they'd have just shut it down, if anybody would have. Because they were just marginal in all those years anyway.

That's just the way it was. Most of them, the Ajax, the Cresson, the big mines, they were all Golden Cycle. And then UGM was owned mostly by Golden Cycle. They had their own stockholders. But, it'd depend on who your shift boss was, or who your Superintendent was. You know, whether you liked the guys and the crews that you worked with. A lot of people changed jobs a lot, but you know, come hunting season a lot of them just quit. They'd just quit for a couple weeks and go hunting, then hire back on somewhere else.

If you quit and went to another, were you working with guys you'd worked with before?

Yeah, oh yeah. Most of them had worked together before. I remember one hunting season, we were on the Strong right there in Victor, for a guy by the name of John. That was an independently owned mine, and he was the one with the money from Denver. A big ol' German guy. Mean guy. So we were working, we were contracting actually, just running the drift. So, we got paid in hunting season. We just took off for three days, come back and he was waiting. He fired us. "I can get more people, I can get more people." Go ahead, get 'em, I don't care.

I went to work somewhere else. He was bad news. That's a mine that could have been shut down for violations of every regulation in the world. They had an underground hoist, their tugger-hoist. It took ya down in the winze that he was sinking on this vein. It was good ore. You had to go down on this bucket, two guys, and you had to take your leg and flip a door to direct the bucket to go where you wanted it to.

See, the bucket's coming straight down the winze. It's on a sheave wheel and this tugger-hoist is letting you down in this bucket. A big round bucket, on a bail. And you're riding the bail, but the winze goes off to the right, see, and then continues to sink. So they had a trap door there. You just had to slow down and kick that door over to where then you were on the guides going down the door to the next 30 or 40 feet. This was on 2600 of the Strong, same as 2600 of the Ajax, they connected right underground. That was a little hairy going down there. If you weren't young and very agile, well, you'd never get down there. (laughter) But you had to have two guys, too. You couldn't ride that thing with one guy, you had to have two to balance it, and then throw the trap door. That was the most accident-prone mine, and I didn't work there long. Six weeks or so, until hunting season come along and it was a good excuse to say "Goodbye."

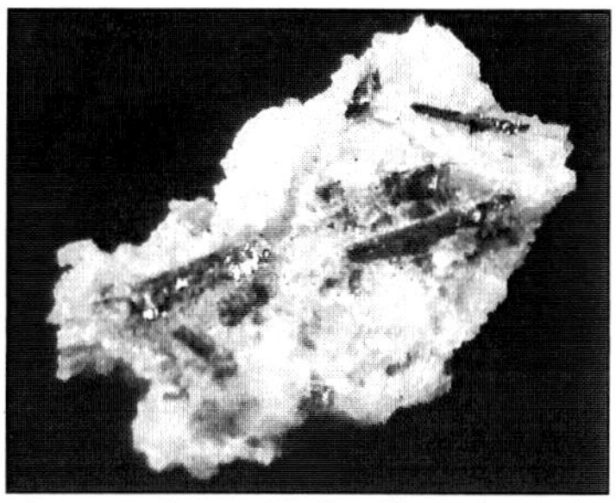

Calaverite specimen Bobtail Vein, Ajax Mine

Communication, Hairnets & White Boots

"It was a very mysterious atmosphere. Dangerous work, and not everybody's going to make it out in one piece." ~ *Brock Gay*

I was born in Leadville. Ten minutes after six, 1950. My dad did a lot of different jobs when I was growing up. He worked at Dairy Farms. He worked the Dauntless Mine, one of his earlier jobs. He worked initiating the first drift down into the Stork Level of the Climax Mine, working a wing-jumbo. Later, after three people on the wing-jumbo were killed on a cave-in, he decided he needed to do something else, and became a welder, working for the Climax Mine. And that's what he stayed in for the rest of his life. He retired after 38 years. Well, about 30 of them, or so, were worked at Climax in Colorado. The rest of the time he worked at their refinery in Iowa. He had high altitude blood disease, and had to go to a lower elevation.

Most of that time, after coming out from the wing-jumbo, he was above ground. He was like working in the mill buildings. He was fabricating. He built, actually, the folding dipper was an idea of one of the miners at Climax. And so, he was in on building the first. And repairing the dippers that would come in that weren't folding. And building cars. He fabricated. When he was 48 he was teaching college in Golden, Colorado. Vocational college, he was like the 'Professor of Welding.' He was an artist; he was very good.

Actually, he did work for an artist. 1962, the Climax went on strike. So Dad couldn't work at the mine, and they prevented him from working anywhere else in the town, you know. If any one of the miners saw him working, then he could get in trouble. So he got a job working in Fairplay, for an artist who would just point where he wanted this metal. My dad would weld it on there, clean it off, and then, Dad told me, he could just go sit in the chair and wait for the artist to make his next decision. (laughter) So yeah, all week long he'd be working for this artist, and come home on the weekend and we'd go fishing or whatever.

The first time I was underground was probably about 1960. I was ten. My grandfather was an avid miner, and he had, he was one of these men who would locate a place, and then he would try and get backers. Or else, he'd use his own money. He'd start working a claim, you know, doing this, doing that to it. A few times he took us underground, just to show us what he was doing. It was interesting, but it wasn't as interesting to me as it was to my brother. When he was 12, which was about 1964, he was in Bonanza, Colorado, with my grandfather. He was actually jack-legging, and nipping. Bringing the powder in, the dynamite, for my grandfather. That kind of stuff. I wasn't really in to that, you know.

How did you end up underground?

I had just gotten out of the military [Vietnam era]. I was so disgusted with the wages for electronics technicians in this country at that time, I came back to Leadville. My dad used his influence, which at the time I wasn't too happy about. But with time, I proved my dad was right. He knew I was going to be a good worker. And they hired me; took me underground and I liked it. It was a very challenging atmosphere. A very mysterious atmosphere, you know. Your light could only go so far, and so high, and so wide. What was farther out than your light, and what was farther to the side of your light?

Communications were more like my personality. Instead of using words, miners' communication is body language and light signals, and hand movements... and the words that were used under there. You could communicate. If somebody was above you and they yelled 'Below,' you knew you better not be there, because something was going to be happening. It was a different language. My first two years were very intense. I was very intense into mining. It's just unfortunate that I was in Climax, Colorado, which isn't a place that they wanted to develop miners. They just wanted a bunch of worker bees. They didn't really care. For example, if you're losing 7 people to death, every year, and you're still considered a good, safe mine, I think, even to this day, that's piss poor.

A lot of times, underground, in an area that wasn't lit, it was only lit by lamps, you just used your lamp for signals. If someone was at a distance away from you, or you were drilling on a machine and you had ear protectors on, they couldn't hear what you were saying. They would signal. They would throw a light out to something. You had to know your position, your place, what you were doing. If you were a chuck-tender, then you knew your job was to tend the chuck. Take care of bits, do that kind of thing. The man who was drilling, he knew when he needed a bit, when he needed a steel to be changed. Every person was trained to do a certain job when you first came underground. After six or eight months, you knew every position.

By using your lights, you could tell somebody to come here, to go there, go to this way, go to that way. I need this over here. Just using your lights, you know. In well-lit areas, with a whole bunch of noise, you could use your hand signals, too. You could tell them to go up, to come down. You could say back this way, back this way (making hand motions). And the old timers, it was old hat to them, but for me it was interesting to learn their language. You never forget, nobody who's ever done it for any length of time ever forgets. After I quit mining I had a headlight that I used at camping, and I caught myself doing that to people. (laughter)

What did you think the first time you went in the dry?

I felt kinda out of place. I was 23 years old, and it seemed like everybody knew what they were doing except me. Of course, I was out of place, because everybody really did. If they didn't know everybody, they'd at least seen everybody else and I was the new face on the block. You know, the older guys were helpful. They'd say, "There's a locker over here you can have." But the ones that were between my age and their early 30s, it seemed like I was some kind of a threat to them. And I didn't realize that they were right, I was a threat to them, just because of the way that mine operated. I was a worker bee and they were not.

The Union had control over the whole mine, and it was a closed shop, you know, in order to work you had to be in the Union. That was the threat that I received after working there about two years. The Union Representative actually threatened to black-ball me if I didn't slow down, because I was making everybody else look bad. It wasn't just me. There were three or four of us guys that we wanted to do the best that we could do. Because they were paying us really good money, really. It was only $6.25 an hour, but in 1973, for a 23 year old, that was good money.

I had to be part of the Union to work there. After threatening to blackball me, basically what I did was bid off to another part of the company, which was several miles away. They were running a water diversion, it was called Ten Mile Tunnel, and I got the bid. There were only 4 of us out there, and the boss. If we did well, the boss looked good, so he didn't care whether we worked 'great,' you know.

So, I spent the majority of my time out there, that I had left at Climax. We went back and forth. We got flooded out a couple of times, had to go back to the main Climax. But they never sent us back to Stork Level. They'd always send us doing details, like salvage work and stuff like that. That was fine; I learned all aspects of mining, by doing all aspects of mining. I wasn't just doing one thing.

My dad worked on the Phillipson Level; I worked on the Phillipson Level. He worked on the Ceresco Level, I worked on Ceresco, too. But I was scrapping them out. My dad was fabricating materials to keep them going, you know. He was interested. He was telling me, "Did you go down to the main shop and take a look at the door there? I built that door." Yeah Dad, we just blew it up and hauled it outside. Worked at the Climax for six and a half years.

What method of blasting did you use?

Sometimes we used blasting caps that were electric. Sometimes we used them that were concussion initiate. They had like a black wick coming out of the cap itself. We'd tie in primer cord and initiate that. Primer cord with a blasting cap, sometimes electric, sometimes just blackwick. We used detonation devices, electrical devices. Some were twist, some were plunger type detonators. Sometimes we'd just use all blackwicks and we'd just tie 'em in with this slow burning fuse line, and we'd just light the fuse and it'd burn through it. It'd light all the blackwicks, and each blackwick we had cut to different lengths so that the burn on the face would go first and make a big hole, so everything else would pull out from there. Sometimes we used non-Ls. They were a kind of a DuPont Canadian device, concussion. They had a small grain stream in a plastic tube that ignited the blasting cap detonator. Hercadets.

Is it true what some say: 'If all safety rules were followed, no mining would get done?'

Well, that's an over simplification. I mean, for the most part, some safety rules are bullshit. They're just there because one time, one guy, this happened to, so now they make this safety rule that limits certain ways of doing things. Not all. But Climax had safety rules like that. They had a safety rules book that was 3" thick. It absolutely impeded work. But the guys got used to it and they loved it. They loved it.

The boss would come tell 'em to do something, and they'd say, "Well, we can't do it, because of this and this and this and this." So they'd just skate. They loved it. A lot of goof-offs. But as far as being safe with handling explosives, Climax was safe. I was in a couple of situations with Climax, though, where they did make mistakes, but really it wasn't their fault.

They got a new explosive in that was a kind of liquid C-4 explosive. C-4's a little bit rubbery; this was more liquidy, so you could actually pour it down holes. You could pressurize it up into holes. Things of that nature. A lot of miners called it donkey dick, but eh, whatever. But it was new, and they put it in a magazine. They laid them one on top of the other, and storing them like that created heat. Apparently if this stuff reached 140 degree temperature, it would ignite.

By the time they found it, the magazine was at 129 degrees. So they had to haul 4700 pounds of this stuff out of the mine and into the valley, back up there. I was recruited to do this, after I had been working at the mine for two weeks. Not knowing anything about anything, really. We went in, they gave us rubber gauntlet gloves, which I thought was to keep the powder from being on top of us, but it was to keep us from getting burnt, picking this stuff up. 7 of us new hires loaded all this stuff up on these boxcars, and they took it out and detonated it in the valley. See, that created another safety rule, but this was a good rule: Never stack it like that, one on top of the other. They had to put pallets between each case.

Could you really tell the personality of the people by how equipment was run?

Yeah, that's true. I could tell who was running the hoist just by the way they would start up, the way they would answer your bells. Blindfolded, or not even looking at the cage, where they stopped ya on each level. Where they landed you. Where they landed when you were caging muck.

You know, you could tell who was doing it, who was running the cage. And some, you could tell what kind of attitude they were in, also. (In a fake accent) "Ah, Butch is running the cage today. Cool. Why are we goin' so slow? Oh, his wife was out, you know. Why are we going so fast? Oh, she really got him good this time." Oh, okay. I could tell who was running. I knew the cager, who was running the cage, even if they switched in mid-shift. It's like who's running the cage now? Because it's not who it was before.

It was the same with trains, too. Muck trains. You knew who's train was being run, when it was coming in. Just by the speed. Like you would be sitting inside a muck drift, or a haulage drift, and you'd be up in the fingers, in the dash, and be filling this train. Well, when you filled that car, you gave him the signal to move ahead. And just by how it was spotted, because some guys, Reuben for example, he would spot the car a little bit ahead so that when you were putting the muck in, you had to make sure the right side of your dipper was a little more full than the left side, to fill it up evenly, and things of that nature. Well, you work with these guys.

Mining, when I was doing it and probably to this day, was one of the most dangerous occupations there was. In the blink of an eye you can be dead. Reuben, for example, was killed. He pulled into a drift and nobody stopped him from going in there, and they shot a stope load and about 5000 pounds of prell in the stope area. He didn't know. No guard was there to guard that entrance. He pulled right in; was waiting for somebody to give him the bells and ba-boom.

There was, at the Climax, there was a safety rule or whatever, that said you couldn't short-wick. The wicks at Climax were set. They were 40 inches long. I wish I could remember what the burn rate was, but it was something like ten seconds per inch. It was quite a long time, really, per inch. They made it so that you couldn't cut that wick down. Because some of these guys that were really in a hurry, you know, they wanted to get things done, especially hang-up men.

You get a big rock comes down and it blocks a finger and you can't get any more muck outta there. These guys were required to get so much muck hauled out to the crusher for the mill. These hang-up men. So what they would do is they would go to this rock and they would put bombs around it, very rarely did they drill it. But they'd put bombs around it, and they'd send down their guards. The guards would block the drifts so that no one could get in. Boom. The rock breaks up, the rock starts falling again, and they're getting their tonnage out.

But, this guy named Butler. He'd been on this crew for about three months. I was working two dashes over doing development work, actually working on developing the dashes. This hang-up man sent out his guards, sent Butler out to one side. He sent out the proper guards. But he cut the wick too short. These guys thought that they had four, five minutes to get out, make their guard stations so nobody could get in.

There were air doors on each side of the drift. And Butler went through the air door, and was closing the air door when the round when off, and the air door smashed him up against the rib. It just... I didn't even recognize his face, and he was still trying to say something, but he was gonna die. They tried to nail that hang-up man for manslaughter, because he definitely broke the rule. He cut the wick in half. See, he was smart. What he did, he cut the wick in half, but not all the way in two. Then he lit it so it burnt this way and this way, so if anybody came back to investigate, the whole wick had been burnt.

There were several though. There was one man I know, he pulled the trains underground. He hauled muck, he'd been doing it for two or three years, and they had what they called pull bottles, and you'd reach out and pull this pull bottle, and throw the track so you could turn this way. And whoever's coming this way can pull the bottle and put the track back. He was reaching out to get the pull bottle and his train derailed, and it ripped off his arm, out of his socket. Knocked if off and ran over it and smashed it. You know, a woman miner, the Climax was one of the first to have a woman. She wasn't a miner, but she was on the muck crew. She came down there and she stuffed some rolls of toilet paper in the socket and wrapped her shirt around like this. The guy's alive. Saw the guy three months ago. I'd never seen so many people up to that point in my life with missing fingers, missing parts of their body, you know, toes.

Did that make you worry about going underground?

No, I didn't worry about it. I mean if I worried about it I wouldn't have made it for any length of time at all. No, I just trained myself to always be aware of everything that was around me. And who I was working with. Because I worked with some people that I just told them, "Go, go, get out of here. You're not going to work with me." There was one guy, Phil, he was just looking to kill somebody or himself. There was just no two ways about it. I worked around him several different places, several different mines, actually, and he'd always show up. He'd always be a major accident. Major accident at Climax. Major accident at the Sherman Mine in Leadville. He worked at the Ajax for a little while until he got hurt there, too.

There was another one, Dennis. He was just a great big guy; he really didn't know how to handle hand tools or motors. Anything outside of his body he didn't know how to handle properly, carefully. Watching out for the other guy, you know. Jack-legs can take... they can do you in right away. Climax had a guy that had been working there about a year, and wouldn't seem like it would kill him, but his hair got caught in the drill and it rolled up and it ripped a chunk of his scalp off. And he died. So then Climax made a safety rule, you had to wear a hairnet. I don't know. They didn't wear the hairnets. It was like, the Union was strong enough to say, "Enough."

I had this friend of mine, Gardener, he was kind of a boss at the Sherman Mine, and his brother was a biology teacher in high school. But his brother didn't want to do that anymore. He was from Ohio. He said he was really interested in mining, in what his brother was doing. So he came out to Leadville and his brother got him a job. I knew one and I met the other one, and I could see why the older brother was a miner. He was big, and there was something that he had. I could see why his younger brother was a teacher, because there was something that he lacked as far as being a miner. We took him underground, and he wanted him to start making good money right away.

Usually in a contract mine situation, if you're a newbie, you start at the bottom. You start as a nipper cleaning out ditches and helping bring steels and putting bits on steel. Well, he moved his brother fairly rapidly, so within a month he was one of what's called Air track driller. It's on tracks and it runs a pretty good size steel. I think it's a 1 3/8" steel and it has couplings. You can drill 120, 160 feet. He had him core drilling, because core drilling, or drilling for fines, actually, you collect fines in a bag. You collect 0-4 feet, and then you drill another 4 feet and collect those fines. 4 feet, 8 feet. We had him doing that, and he was out in front of the rig, behind the bit, changing a steel. Started it up, and the steel got hold of his wet coat, slicker, the yellow slicker. And wound him up so tight it crushed his chest and basically he died of suffocation.

He probably would have been able to be saved if somebody had been there within two or three minutes. In that mine, see, that was... Climax had a safety rule that said wherever there's one miner, there's two miners. And never will there be one miner. At this mine, it was like, "You feel safe alone?" "Yeah." "Okay." Most of us worked alone. I had five drifts, and I had five machines. I would see a guy at noon, and at quitting time. The only other time I may see somebody is if I went to lunch. Then I might see two or three other guys, but basically you were on your own. But he was found half an hour, forty-five minutes after the accident. His brother, he was devastated. He was just devastated.

I fell 250 feet out of a raise once. I was way up there in what they call a dog-hole raise, it's only this big. I was just following a vein of silver, all they wanted me to do, all I was getting paid for was to get the silver. So some places it'd be this big, some places it'd be just enough to get you through. The stull that I had across there, the rock on both sides, just went out. What helped me was I had a bunch of bracing and stulls going up there and I was grabbing those on my way down. My machine was going down before me, so it kind of cleared a way.

When we hit down there, the machine went to one side, and I went to one side. The funny thing is when I came down, one of my buddies was just coming in from the muck pile. "Oh, I'm glad you came down to see me." (laughter) "Where's my hardhat? Where's my lamp, where's my hardhat?" That's the first thing I said.

At the Climax I got buried twice. Doing some things that were just... we needed to get this done, we needed to get these timber sets in here. We needed to stop this ground from falling in. It was right on the Mosquito Fault, and it had already peeled up about twelve to fifteen feet. Nobody, all the sudden, none of the older miners knew anything about timbering. "No, I've never timbered before."

So I got some diagrams from some of the old timers. They showed me how to make a Dap set. You just cut the post at an angel like this, then you cut a notch out of the cap like that, so everything locks in. One of them actually had this book on mine sets. He said, "Use this." So I ordered the timber, 8x8"s, 3x12"s, and all the knee bracing material, 4x6"s, you know, everything it said in this book. Because I hadn't done it before.

I just started out, put the first set in, ran stringers out, put the next set in, run the stringers out. In the meantime I'm putting in foot blocks and footbrakes, knee braces, collar braces and all that kind of stuff. One guy finally decided, once I had three sets in, that he was going to give me a hand. He walked out and he was tightening everything up. I got out in front of the last set, I had some stringers out, and it just paid off. The stringers just went like this, basically what it did was broke everything up so it was nicely stacked.

I was standing there like this, buried, with six guys standing right there looking at me. Well, you know, "Get out here!" Finally the oldest guy, he was 67 years old, cussed at them and started pulling rocks and stuff off of me. I just continued, I laid the timber right in the muck and just continued on. Instead of mucking it up, I put foot blocks in. Said the hell with it, this is how it's going to be. Mosquito fault is a fault. It's mostly talc and pyrite cubes and crap like that. Along the edges the rocks are kind of deteriorated, as you get towards the middle of it, it's like this huge sticky stuff, and full of pyrite cubes and that kind of stuff. But once it starts falling in, shew, it could go all the way to the surface.

What was your first experience mining - what did you do?

My first day at work, I basically just walked around. Took 2x12"s off of the drainage ditches, which they didn't tell me were called drainage ditches, they just said, "Clean these piss ditches." And so we shoveled up what was collected inside them into a ten ton car, which of course meant shoveling it up over our heads, and I was with seven other guys who had been hired that same day. We just wandered around underground. I didn't know where I was at, I got turned around.

There was one guy that was kind of like Climax's biggest guy, he'd been working there for 30 years or something, I don't know. But he said Okay, come on, we're going to the 3200. "Where am I now?" (laughter) At the end of the day he took us back, I'll never forget, to the mantrip. It was such a strange feeling, walking around in the dark all day, and then coming out into the dry. At that time it was probably 400-500 guys per shift, because Climax had, I think, a little over 1200 people working there. By the time I quit they had a little over 3800 people, I think it was, working there. It was getting crazy. They had one boss for every three guys. Climax couldn't deal with the Union. Instead of forcing each man to be more efficient and actually do work, they put the pressure on these supervisors to watch over these three guys or whatever. I don't know, it was a strange mine.

My dad loved it, but that was because my dad only had one real job in his life. I mean from the time he worked at Climax until the time he retired, there was no break in service. So he had loyalty for that company. Dad and I never talked about Climax, because he knew how I felt about the place. But then he reconciled it with himself and with me by saying, "Oh, I work at the mill, you work at the mine."

What do you think of Tommyknockers?

Well, I myself, I've heard strange things. But I have explanations for what it was. But I've heard all kinds of things, like Climax has this thing called 'White Boots.' Supposedly, and I'm sure there's other people that could tell it better, a new guy came to work and he was wearing white mine boots. He was killed, and when they dug him up the boots were gone, and now they're roaming around. When I was at Climax there were definitely miners that were scared to death. You mention 'White Boots' and they'd look for the nearest exit out of there.

Did they hear him walking?

I guess. I heard some walking noises when I worked on the C Lateral. I heard like boot splashing noises and looked around with my light, but there was nobody there. But you know, the guy that was with me, he heard it, too. I looked for it and he asked me and it was like no, I didn't see anybody. Where'd he go? That guy was out of that lateral. He was gone.

There's a lot of superstition in miners, I found. Like one guy had a lucky pocket watch. One day he forgot it, and he left from underground to go get it. He'd left it in the dry. He'd had it with him for like four years. Been mining for four years. He wouldn't go underground without that watch, that pocket watch. And it didn't even work. Didn't work. Yeah, they're careful about what they put on first, what they take off first. There were a lot of guys who were kind of leery about animals, like mice.

Do animals wander into the mines?

I've seen porcupine and bear. A porcupine or bear won't ride the cage down to a level and get off; a mouse would. There's mice in shaft mines, sometimes even rats. A drift mine that's open, you know, an adit, bears would go inside one of those just to look around. I mean this bear had gone in and turned left, he was in the mechanics shop. Nice warm area with a trash can and all that kind of stuff. So you know, there've been a few sightings like that. But what's kind of neat about virtually every mine I've walked in to, you'll see a mouse coming down the track. Just following the rail in. (laughter) Hit him with the light and he'll stop, then he'll just keep right on, keep comin' right in, you know. Even in the Carlton Tunnel, you know, the time that we had to evacuate through the Carlton Tunnel because they had a cage mishap in the Ajax, there were all kinds of animals at the adit area. Bird nests. Because of the warmth coming out of there. Of course, there's that stream going out of there, too.

Why did you have to evacuate the Ajax?

Well, a piece of material that they were bringing up fell outside the cage and grabbed a hold of the shaft timbers, and it ripped up a bunch of shaft timbers and hung up the cage. So, we had to send guys down from the top to about, I can't remember, I think it was about 800 level and they had to go out and break lose the cage. Re-timber it.

How did you find your way to the Carlton?

Well, you just go up to the 3100 level and it comes off, I think it's the 3100 level. If you sent me back down the Ajax, I'd find my way around in a couple of days. I think the Vindicator has a tunnel out to the surface, and access to the Vindicator is on the 2100 level of the Ajax mine. But I think that the Carlton is on the 3100 level. I mean, all we did was we just climbed up approximately 300 feet to the 3100 level and just started walking out. It's not hard to find your way, but it was running a lot of water. We found out that there was a lot of mud.

What did you think of your first day at the Ajax, after having been at the Climax?

I was pretty rowdy by the time I had worked at the Sherman, contracting for two years. That's what I did after the Climax. I'd done some other little jobs for some farmers, using explosives. I worked with another guy that did a couple of the ski runs at Vail. I worked for a guy and did a couple foundations for houses inside the City Limits where he couldn't break up any glass or anything. He was very, very good. Explosive demolition expert, is what I'd call him.

So, I was pretty rowdy when I got there. I was like, "So what?" They asked me at first, "Have you ever been in a shaft mine?" I told them, "Yeah, but I've never ridden a cage." Because I hadn't. When I was a kid, we'd climb down the cribbing. At the Climax we went in on a train. Sherman mine we went in on a train.

What did you think the first time on the cage?

Well, it was okay. It was fun. You know, when you ride in an elevator, you don't see the walls going by. That was my first real thought, was nothing outside the cage. Just keep everything inside the cage. So when six or eight of us hopped on the cage, I just made sure that I was kind of in the middle of the group, you know. And then the ride was forever. It was like fifteen seconds, one minute, two minutes, you know, just going by like that. But as far as the dry and the mine and all that, I didn't get the feeling like the first time at Climax, and going underground. It's the same except one your accessing straight on, one you're accessing straight down. The 3350 level was miserable. It was constantly wet, very warm. Just a miserable place to work. So, I just coped with it by working.

Was it dry in Leadville?

It was wet in some places, but you never had that constant on your head, in your face type wet. It was wet on the drift, maybe leaking out of the sides of the ribs on the walls. Ten Mile Tunnel was wet for about 500 feet, but there was a reason for that... just as there was a reason, I guess, that it was wet on the 3350. (chuckle) It's just groundwater leaking down through all the cracks. But if the water would have been as warm as the air temperature, it would have been a little bit different. But the water wasn't warm.

Were there pranks underground?

There were quite a few pranks. I didn't pull any pranks, because I always thought it was too dangerous. If you're doing a dangerous job, don't make it more dangerous by being stupid. I remember in that same C Lateral in the Climax, these two guys took some mine boots and painted 'em white and stuck these bib overalls with stuffing, put a hard hat on it, and put it down the C Lateral, knowing this guy I was telling you about was going to go down there. He ended up going to the hospital. I mean, he ran into a huge timber and his heart started fibrillating. So here's these two guys that were just screwing off at work, and they injure somebody who was there to really do some work.

And another guy, when we were doing salvage on the Ceresco Level. I don't know if you've ever seen these 'redhead guns' that shoot connectors into concrete and into wood. Call them redheads. They use bullets like .22 caliber, .38 caliber. Well, there was a whole shelf of these blanks that basically, they're just blanks. We were supposed to just salvage all the rail and materials and everything. Get them to the surface so they could either reuse them, or put them in the salvage yard.

Well, he took the powder out of these things, and it was around the 4th of July, so he had firecrackers, and he stuffed all this black powder and bolts and nuts and stuff in this pipe and blocked the end of it. He put a fire cracker in there as an igniter, and he put this pipe in a vice inside of this shop area that we were supposed to be salvaging. There were two other guys with him. And he says okay, everybody stand back here behind me, we're going to shoot out the door.

He lit that thing off, and I was about 4 drifts away and I hear this "Boom!" I knew that there was nobody, there was no blasting going on, so I headed towards that area. Pretty soon I get to a drift where a lot of dust is blowing out, so I know the explosion's down there somewhere. I put on my respirator and started walking down there and I see this piece of wood hanging down on the rails like this. Looking a little bit closer I realize oh, that's the shop.

Anyway, when I get to the shop one guy's just coming out and he's going 'Ahh.' "What happened?" He says, "I don't know. Something exploded." I went in and this guy had one of those bolts... the whole piece of pipe, you know by now, the whole piece of pipe just blew away. The bolts went through his knee and he was laying there, he was out cold. I thought, oh, there was something a lot more wrong with him. I called outside. I told the guy to take the motor out and bring in the basket. We didn't have one underground because we were doing salvage. And he brought that back in.

By then Safety is notified, hospital's notified. A whole investigation goes on about this whole thing. Everybody that was on that level that day got cited for some kind of violation. Two guys got fired. One guy got 45 days off, without pay. I got 30 days off. Then they changed that to 5 days off, when they realized. But they still kept me being penalized because I was on that level. They considered me as the supervisor in charge. In actuality, I was a day rate supervisor and the supervisor was actually there at work that day. But, that's what goofing off at work can do if you're doing it in a dangerous job. It's like, say you're a fireman. Somebody greases that fire pole, and you hurt two firefighters in the process.

When you worked in this District did you ever see or hear of walking underground between towns?

I've seen the Vindicator, the Cresson, the Mollie Kathleen, and I've never entered them through the top. I don't know if I could follow that path again, but I was with someone, and he had gotten the information. I don't know who he got it from, to tell you the truth. To get in to the Vindicator, there's a gate on the, I believe it's the 21st level, and that's where you start. And you have to go through the Vindicator, and it's connected to the Cresson. It was when I'd been underground there maybe six months. We never came up anywhere, we just moved through.

We found unused explosives and machines still in one side of the Cresson. Scary. Then he said, "I think this is the Mollie Kathleen. Let's get back." Who knows, it might not have been that far, because that's like seven miles or something. But the Cresson is only maybe a mile and a half, two miles at the most. I do know that you can tell every time you cross a mining property under there. There's some kind of, whether it be a gate, whether it be a difference in the level, like you come out and have to go down, or you'll come out and have to go up. You can move across a whole bunch of property down there. It's not all caved in. I think Leadville's pretty much caved in.

Cripple Creek-Victor has harder rock?

Different rock. Different structure.

Did you do timbering?

I didn't do timbering here. The only timbering I did was when we took the shaft down another 65 or 70 feet. There was timbering in a winze, and we did timbering for our ore chutes.

How did you go about making the shaft deeper?

We put a pump down in there. We didn't do much at the Ajax. We just kind of finished up the 3350 level and went down so we had a place where we could drop the cage down below the 3350 level safely. Whereas before, they'd drop it down and it was right above the level and you'd jump outta the cage. I did a lot more shafting at the Hopemore in Leadville. We basically did things the same way. Used a clam, strapped it to the timber, and we pulled out the bottom. Drilled down holes, blow 'em out. Put our pattern in, ride up on the cage. At the Hopemore we rode up on the cage. At the Ajax we'd go to 3100 level, get off, walk back. Because it was safe enough. But the Hopemore, there was nowhere to go. You didn't have a level that you could get out on, so they pulled us all the way to the surface. Then you'd go down and you'd have that clam, just drop it down.

What were you taught to call the frame that sits over the shaft?

The headframe. I mean, I called them gallas frames until I started mining. Because that's what I was told they were called. That's what I called them when I was growing up. Gallas frames, yeah. When I was young. When we were hanging around, my mom would say, "I don't want you kids hanging around those gallas frames." (laughter) "Stay away from those shafts."

Why did you start calling them headframes?

Because I went to mining here, at the Ajax. At Climax, they had a shaft that had a hoist that took you down about 1000 feet. It was totally enclosed, it was inside of the building. You didn't see it. You didn't see cables, it was more like an elevator. A huge elevator. Huge. They could send down front-end loaders. They would take the dipper part off of it and put that in there, but it was pretty good size. Here it's headframe. Apparently Gallows was the one that designed the frame with the wheels and the balance that you needed to have. And the distances back for the depth that you were going. As far as I know, nobody was ever hung on a 'headframe.'

Tell me about high-grading?

High-grading's never been an issue in any mine that I've ever been at, except the Hopemore in Leadville. The Hopemore in Leadville had nothing worth high-grading. Which that seems so strange to me.

The only thing the Climax didn't want you to high-grade was their tools and materials. The whole thing changed when, at Climax, the guy that I was working with and two other guys, we went into the Mosquito fault and found brick sized, and ten inch sized cubes of pyrite. Real nice looking stuff. I wasn't about to just send them in to the mill, so I was taking them out, cleaning them up and stacking them back in the dash we were working. We hid them in the back of the dash.

It just so happened that this superintendent and the mine foreman and our shift boss came into that dash, because they wanted to see the kind of work we were doing. I was telling you we were having problems because we were actually working. They saw this pyrite and actually the mine foreman asked us if that was ours. And no, we just don't want to send this to the mill. He says, "Well, do you mind if I take one?" But after that they put out this paper that said nothing, property wise, including minerals, are to leave the property of Climax in somebody else's possession.

It was right after that they hit a seam of rhodochrosite. The thing that was really sad about that is that Climax had just put out this high-grading thing, and to make sure that it was abided by, they sent down two, one engineer and a geologist, they high-graded the rhodochrosite, and the rest of it was sent to the crusher. Of course, that's waste then. It's out there on the tailings, just rhodochrosite sand.

But at the Ajax, there was no stipulation there against high-grading. They hadn't found anything there really worth looking at for years. I don't know how many years before. I know it was at least four years before I got there. You know, pyrite, that was about it. The 400 level had real nice chunks of pyrite, and 800 level, I think it was, had some nice fluorite. You know, there was some nice dolomite shale, good samples of that.

But nothing, nothing like what was found on the 3350 level. You know, when we ran into that, it was obviously what I was bringing up from underground. Nobody else wanted to take any of it up, because they thought it was worthless.

When I brought up Sylvanite, the first Sylvanite that I found in there, you know a rock about that long and 3" thick and a foot wide. I mean, I carried it out and I took it to Ed Hunter and said, "What do you think this is?" He says, "Well, can you leave it here?" Well, yeah. He said looks like Sylvanite, where'd you get it? He said the engineer thinks that it may be

Sylvanite, but it's probably pyrite. I still have a piece of that, and it's Sylvanite.

When we started developing the drift, that's when the word got out that there was some pretty high-grade stuff down there, and that's when some other people started getting involved. It turned into a kind of nasty little scene where we would leave at the end of our shift, and when we'd come back the next day we'd see where somebody'd been there and they'd been slabbing off the rounds and taking rock and stuff. I'd always ask the shift boss, "Who comes down?" So, it was kind of a strange situation; I don't know what was going on.

I do know that they had estimated that there was approximately ten million ounces of gold in that vein. You know how they do a reserve estimation. Well, Hecla said there was nothing there worth mining. The thing that was sad about that situation is all of the development, we had just completed all of the development on six or eight chutes, and the stope, the shrink stope had already been started. And the chutes, all that was left was to shoot rounds and pull, pull, pull. See, that's what was going on, we'd go underground and a chute would be pulled and empty. Where'd it go? Where did it go? Yeah. That's a Tommyknocker for you.

Missed rounds are the real danger?

Oh yes. We called them "Assholes." Well, actually, not really, it's a missed fire. Asshole is like you drill a six foot hole and you leave about this much of the hole still in the rock. Your objective is to pull it all, if you drill six feet, you want to pull six feet all the way. A miss fire is one that's just hanging there. It's got the cap in it, the dynamite or the prell in it. Those are kind of dangerous. It's best to wait like 30 minutes to an hour after you shoot a round before you go back in. That gives it time for any slow fire, or something like that to happen. I pulled miss fires before. Just clean them up, clean the powder out, and take the cap out. The next round put it back in and shoot it up.

You really have to watch for those, especially when you're drilling. You don't want to drill into one. That's why cleaning the face off, so you can see the whole face... I knew miners that would just run right in. After they shot a round, they'd pull it so the muck would be spread out on the whole drift, so they could get back in and start drilling again, right away. They'd just put their respirator on. But that's too dangerous. Too stupid.

I've done everything from hand-steeling, real hand-steeling to shoot rounds, all the way to running a triple boom jumbo, and it's crazy the kinds of equipment they use underground. It's getting safer, I think, except in the coal industry. I think the coal industry is still as bad as it was before, even with the improvements in the equipment.

But with the sophistication and equipment they have now, underground mining is probably kaput. It will be a small industry, but it won't be like it was back in the 30s, 40s, 50s, 60s, 70s, 80s started dying off, and now it's open pit. Just come from the top. I'm not sad about it. I've had some good experiences. You're never going to get rid of underground mining. There's always going to be somebody that's going to hunt something that's underground, and that's going to be the way to access it. The best, least costly, and perhaps even the least dangerous way to do it.

Some fine, modern claim jumping happened in the '80s in Cripple Creek. There were people jumping onto claims with turquoise. There were people jumping on a claim up at the Ajax, I know that was going on. There were people back up here jumping on a gold claim. That sea foam turquoise. It was like the old days, you know.

Were things run similarly at the Ajax as at the Climax?

No. What was the basic difference, I think, was the experience at the Ajax as far as the people that were working there. They wanted to learn what was going on, and how to make it better. There was always this belief that there was going to be a rich strike. The next Mother Lode was coming in. Climax, you were just mining molybdenum. You were just tearing up the earth and throwing it in the mill and crushing it.

These guys in Cripple Creek, we were all looking for a strike. Not so we would be rich, just so we would be able to see it. Climax was just a machine tearing down the earth. Ajax was like this pointed search for the Mother Lode. That was the whole goal. The goal was to get all this waste rock out of the way so we could see the gold.

Climax, all of the rock at Climax has Moly in it, but most of the guys up there didn't even know what it looked like. Didn't matter. The Climax everything went to the mill. They had a huge, huge crusher on the 600 level. They had a belt that went from the 600 level to the Stork level. They had a huge mill on the Stork level. They had a belt that went from Stork level all the way up to another area where there was a huge mill above ground. And they had packing. My Dad worked there, he knew what it was called. I just called it a packing plant. To refine dried molybdenum, put it inside these fifty-five gallon barrels and it was an operation. An operation.

One of the jobs my dad had was to take a dip out of the smelting and put it in an assay. I don't know exactly, he used to take the dip, or the guy he was working with took it. Used an old cigar lid, and dipped one of those at each smelting. Take it and dump it in a can. My dad and he were pretty good friends, and he decided he was going to smelt it down.

Well, when he smelted it down it came out to 50 ounces of gold. And at that time, people weren't allowed to have gold, let alone 50 ounces of gold. So, they took it back on the swing shift, they were working day shift, they took it back on swing shift because they knew nobody would be in there. They put it back in the smelt. My dad said if he'd been older than 17, that would have never gone back in. 50 ounces in 1980, 50 times $800, that'd be a lot of money. That's when silver was I think about $60 an ounce. One guy that was working up there at the Sherman as a contact miner, he was going to make all kinds of money on silver. He was making about $6,000 a month, and pumping about $4,000 a month into silver. After it turned $40 an ounce is when he put most of his money into it. It went just like the gold did, just even faster.

Theresa Mine & Pikes Peak

The Skipper, The Mule Skinner and The Dump Rat

"When I was eleven years old, I started working on the top car on the Vindicator." ~ Michael 'Mickey' Nothaus

I was born in '41, at Cripple Creek Hospital. My grandparents, John and Ida Reid, had a farm in Missouri in the late 1800s and he couldn't make a go of it. So, he moved here to Colorado, hoping he could find a job working around the mines and things. They moved here and he got a job with the City of Goldfield, as the Marshall. They moved to Goldfield and they lived just above the South School, where his wife Ida was a teacher. Granddad kind of worked around as Marshall, and also part-time as a dump rat.

A dump rat is where they've already brought the ore out; they used to have certain guys would get a lease on a dump. In the early days they brought the gold out, they only shipped exactly what was really good ore. The rest of it they dumped out as waste. These dump rats would get a lease on these dumps and then they'd work it. They'd sort that good ore, and the bad ore out, and they would make money. They would ship it that way. He had a dump right on the Vindicator there. When I was growin' up he was still working that old dump. He got a lease on it, he had a big bin out there and they'd dump so much ore rock into it. He'd get that thing full and bring a railcar and put underneath it. Load up this railcar and send it to Colorado Springs. He made a little money by doing that, on the side.

During the early days, oh, when they blew up the Independence Depot, well, they asked my granddad to come to Victor and help with crowd control. This was the day after. And he was ridin' a horse there, oh, right at Fourth and Diamond, by the Union Hall, and some sniper shot him in the back. So, they shipped him by rail to St. Joseph Hospital in Denver.

He was there, and he had gangrene in his leg, so they were going to let him die. He said, "Send me back to Cripple Creek. They'll cut my leg off up there."

So, they sent him here to Cripple Creek and cut his leg off oh, about six inches above the knee. He got out of hospital, he got him a piece of plywood so thick and he got some pipe fittings, and he built this steel peg leg out of these pipe fittings. He attached it to that piece of plywood, and he put straps on it; strapped it up. He had a steel peg leg. They say he was one of the first people to ever have a steel peg leg. It was quite an impression at that time. He walked around Cripple Creek and Victor 50 years after that. (laughter)

My mother Portia Reed was a school teacher in Missouri. My dad, Albert, worked on the farm there and he'd get up early in the morning and build fires in that school for them. That's how he met my mother. He got tired of working on the farm all the time and she said, "Well, why don't you come out to Colorado and you can get a job out there." He worked with my granddad, up there on that dump and the people at the Vindicator found out that he could handle mules. He was a mule-skinner, because on the farm he had these mules.

So, he went up, and went to work there trammin' underground with these mules, at the Vindicator. He done that for years. Those mules were really, really smart animals. Even though they were blind from bein' underground, they were smart. You had to give 'em a cigarette in the mornin'. They wouldn't smoke it, but they would chew that cigarette. You couldn't get 'em to move unless they had that cigarette. They'd pull four cars. You put a fifth one on there, they wouldn't move. And when they put 'em down there they had carbide lights on their harnesses. They had to have that carbide light on there, or they wouldn't move. Didn't have to be lit, but it had to be on there. They could go down those tracks and they could step in between those ties just perfectly. If there was some gas or anything, walk back in there with them and their ol' ears would go back and they'd back up. 'Better get outta there, because that's full of gas.'

If he wasn't out on the station at lunchtime, they were. They had pretty much underground little barns built for these mules here, you know. Boy, they were up there and ready for their lunch. After lunch, give 'em another cigarette and they'd get to work. My dad took those four cars, and built side boards on them. They didn't know the difference, so he outfoxed them on that. But that's all they would haul, is those four cars. They were pretty smart animals, for bein' blind.

I remember when they took 'em outta there. I wasn't too old then. Well, I was, too, because I'd been workin' up here at the Vindicator, off and on. My dad give up the mules and he went to leasin'. He was minin' with a couple of other old guys from around here; he went with them on a lease. They would bring the ore up during the day and they'd dump it in the ore house out there. Then when I got out of school at night, I was eight years old, I think, I'd go up there and sort ore for my dad. Then when I was eleven years old, I started working on the top car on the Vindicator. I was still goin' to school, but they was working two shifts up there, and I run a car at night.

The underground cars bring the ore up and they dump it into this bin, and then they go into the top car. If it was waste, take it out and dump it on the waste dump. If it wasn't waste, then they'd take it out and dump it into the ore house, that had certain bins, at the ore house, it had to go in to. The skip would come up... you had to be watching that skip because it was numbered like waste or bin and it had one, two, three, four. They had this bar on it, as it went kinda by you had to catch the number on it so you know which one of the bins to dump it into out there in the ore house.

Well, it was quite a job. There were a few times that the hoistman, he wouldn't be watching what he was doin' or I might be still out there in the ore house or something, didn't have my car under the chute. He dumped it on the deck. He'd have to come out and help me shovel it up in there. Oh, I musta' been about maybe thirteen or fourteen then. I was eleven when I started runnin' that top car. Then I'd go to school during the daytime, you know. Sometimes I found myself comin' and goin'. (laughter)

When I was fifteen I went underground. That's when they took the mules out. Took the mules up from underground there on the Vindicator, and course they had to destroy them. They bought these air-trams. Most of them were built in the blacksmith's shop over on the Cresson. That was my first job underground, was runnin' an air-tram. It had a big tank on it; you'd fill it up full of air. It had an air motor on it. Just opened up the air on it and go. Every so often you'd run low on air, well, they had air piped back for drilling in there. You'd just pull it in there and fill up the tank and on down the drift. They were pretty efficient. Kind of dangerous in a lot of ways because you had to sit sideways on it, and try to watch both ways. Well, sometimes those chutes would be pretty low and as you went under them you had to lay down there and let these chutes go over you. That's the way they loaded the cars. You had to really be watchin' because a lot of times, if you'd forget what you were doin', could get your head knocked off. That was my experience runnin' the air-trams up there.

Then they closed the Vindicator down, and my dad and I went to work on the Ajax. He worked for the company, and I also worked for the company, but I was a skipper. That's the guy that took care of the skip. Like if they had dynamite and stuff to take down, they'd take it down. If they had pipe or anything to go underground, they would. Then underground, when the skip come down, the bucket would tip down and he would dump the tram cars into that bucket and send them back to the top. That was a skipper, that's what they called them. I worked that. I worked in the ore house first, before I went skippin' on the Ajax. But I couldn't make enough money in the ore house up there, sortin' ore, so I got this job as skipper.

You were in charge of the stuff coming in and out of the skip?

Right, yeah. We took all material down, and everything else out. The hoistman, who was on top all the time, we talked with each other through the bell system. And you know, like Gene Huber was one, we worked like on Saturdays and Sundays and we'd run the shaft. And the skippers and the hoistman would be there. Well, I was givin' Gene a bad time. That was the worst thing I could'a done, because I had to go down to the 2800 level of the Ajax and get some stuff, before we started runnin' the shaft. I climbed on that skip and he dropped it. I mean he dropped that thing. I was clear at the bottom of the skip and you go 2800 feet and it was just like free falling, and just before I got to 2800, he catches it. Man, it just 'Boom, Boom', up and down there. The skip had a little hole in there to let the water off. A couple times I thought I was gonna go through that little hole. I didn't get hurt, but I knew better than to give the hoistman a bad time after that.

We worked together for quite a few years at it, doin' that. Working with Wilfred Bielz. We hadn't had a day off in probably six months. We worked all week and run the shaft on weekends. So we decided we were gonna go get drunk. We went out and got drunk, come back, and the Superintendent says, "Well, you guys are gonna give me ten days for not showin' up for that Monday for work." So Wilfred went to work for the State Highway Department in Denver, and I went to work for the railroad in Denver.

I didn't pick up minin' again 'til later on, 'til I quit the railroad after 24 years. I come back up here and went minin'. And I worked on the Mollie Kathleen, herdin' 'turkeys' through the Mollie Kathleen. I enjoyed it.

Then my shirttail relations wanted me to go to work down here on the Sleepy Jim. It's south of Cripple Creek, there on Shelf Road. "Oh, all ya have to do is show us what to do." And all this stuff. Well, that don't work out as you're workin' in there. Instead of a vertical shaft we were runnin' an incline shaft, which went down gradually. What they wanted to do was go around and circle. There was a place there that had pretty good ore. I figured it had good ore. Old Jim Young showed me, one day we were down there messin' around, he said, "If you keep goin' toward that area right out there... see that spot up there on the hill?" and I says, "Yeah." He says, "It never has any snow on it, it's always clear of snow." I got to watchin' and sure enough.

Well, that's what we were headed for was that field. One day it'd be really good and the next day they had nothin'. It was just like the volcano came over and hit this granite fault and then folded back under. Loses all pretty much track of your fractured rock in there. So we were workin' in there one day and I shot a round and the top holes didn't go. So, I'm sittin' there trying to figure out, you know, how to get those top holes shot. My two partners went on out to get more dynamite and I'm sittin' there figurin', and I felt rock come down. Hit me on the head and the back and I figured well, I better get out of here. I started to turn around and I hear a big old slab of rock hit that timber, and took out the timber. I was under it and the rock. They estimated there was over 5 ton of rock and timber that I was buried under.

My face was in my hardhat. I could hear, kind of, what was going on, a lot of it. I heard them say, "We gotta get him outta here, because it looks like some more's gonna be comin' down." So I think I started pushing on it, too. I think I broke my back that way. They dug me out, flew me out by helicopter to Colorado Springs. Well, they said there was nothin' wrong with my back, they couldn't find anything in x-rays. Pretty soon they decided to take an MRI of my back and found it was broke in three places. When I got hit, my spine, instead of bein' straight, it curved. Where that curve was is where it was broke, so they had to go in there and try and straighten up that spine. Had four operations on that spine. But it was never right. I guess I got off on the wrong track.

Was that the end of your mining?

Yeah, pretty much. I went back to the Mollie Kathleen; try to take some tours through the Mollie Kathleen. Kind of weird to be down there, you get this awful weird feeling like the rocks are going to come in on you.

How long were you buried?

It seemed like forever, but they said it was only about fifteen minutes. It seemed like it was longer than that. Where my legs... one of my legs was over one of the timbers. It broke my legs, broke my pelvis. I was pretty well tore up. They were worried that I'd never walk again. But they done that surgery, at least keepin' me to where I can walk now. They wanted to do more surgery, but I said no, that's it. They fused my back there, and the fusion didn't hold. So then they take part of your hip, to fuse your back. And when it didn't hold, then they had to take some more outta this side. Now I've got arthritis in those hips, too. (chuckle)

Did they keep mining at that mine?

No, well, Eddie helped them for awhile. To this day, I don't know why they ran that drift. We were goin' this way, and they wanted to run a slusher drift out this way, and there was nothin' out there but that granite fault. If they'd 'a kept goin' another 75 foot, they'd 'a hit that hot spot back there. Had to be somethin' comin' up through there because that ground was warm all the time. Some kind of heat comin' up through there. You know, I can still go down there in the wintertime, I can pick that spot out. It's still dry. A lot of times you go down there early in the morning and the deer will be layin' on it. It stays pretty much green in there. But it's somethin' if I was gonna do it again, I'd go up there and drill that on a

45 degree angle, a couple 'a different ways. Get a core sample and see what ya got in there.

When was the very first time you went underground?

The first time I must'a been probably eight or nine, because I was workin' up there with my dad. My dad took me underground then. He had to come up on top to get some stuff and he said, "You wanna come down for a while?" I said, "Sure." So we went down underground. That was the first time I remember. It was on the Vindicator.

I worked a while on the ore house over there on the Cresson. Cresson, Ajax and the Vindicator. I think the Vindicator's where I spent most of my time. Growin' up there, Dad was workin' there. A lot of things went on; that was a big mine at one time, the Vindicator.

Victor used to get water out of the Vindicator, for drinking. They pumped something like 28,000 gallons a day to Victor, outta that. A lot of people don't know that. See, they had the Carlton Tunnel run up to the Portland, but it didn't get over there to the Vindicator. Before I left they were talkin' about runnin' a lateral over to the Vindicator, and mining up to the 3100 foot of the Ajax. Well, the way I understood it they run it out so far and the geologist or whoever was the surveyors got it turned and they went over to Goldfield there, and hit this Goldfield Fault. 'Cause there's a fault line that runs all the way down, from the top of Victor Pass, right down by the Theresa, all the way down through Victor. And one side of it, if you go down through that, you'll notice there's not very many good mines over at Goldfield there. They were all on this side. Well, that fault kept that ore cut off. They was runnin' that lateral, they ended up in that Goldfield Fault and far as I know, that's as far as they went. I don't think they ever got over to the Vindicator. But there was good ore on the Vindicator. They're minin' it today. Used to be around Hull City and the Vindicator and everything over there. They're not very far from there. They had good ore at the Deadwood. That's what they're findin' today up here. I look up there and it scares me. Gettin' closer and goin' down farther each time. I look for Independence and all those mines to be gone before long.

Were there lots of accidents at the mines when you were a kid?

Not very many. Not as many as what they thought. Sometimes, you know... back in the early, early days, yeah. But then mine safety, they got to where they stepped in. They'd drill dry and get silicosis or that dirt on the lungs and things. And they made 'em start using water when they were drillin'. Their explosives were so much better later on than what they were originally. Dynamite was hard to handle. It would go off when you didn't want it to and everything else. That's where most of the miners got hurt, was that. I guess too much nitro in the dynamite, and if you shook it wrong or done somethin' wrong with it. Then this newer dynamite they came out with where it took six hundred pounds of pressure to set it off, that made it a lot safer. Use dynamite caps to set it off. Now today they've got so many different kinds of explosives, it's serious. Everything's electric. Like at the mine, they don't use dynamite, they use prell, which is that fertilizer and diesel fuel, is what it is. It's maybe not as strong as dynamite, but the way they do it, you know, they use that diesel fuel and then they set it off. They don't want it to throw rock all over. It just picks it up and sets it right back down. That's all they want, just to break it up. So, it works out real good for them.

Do certain mines have better ore?

Yeah, actually they can't... that leaching process doesn't touch that Sylvanite or Calaverite. It doesn't touch it. So they're losing all that. But most of what they're mining is free gold. The way I understand bein' free gold is that when the volcano started to erupt, the gasses came out and like in California, they erupted, they spewed these gold nuggets all over. Well, up here they didn't really erupt. Pisgah was one of 'em, Nipple Mountain, that were actually volcanoes that erupted. The rest of these are bubbles that just kind of bubbled up and they cooled down. As they cooled down the gasses that were in these cracks and crevices of the earth cooled down and caused this Sylvanite and gold to mix. Oh say, below 400 foot. After four hundred foot, you get mostly Sylvanite and Calaverite. Above that it had already turned into gold. It had melted. That's what they're minin' now is that free gold. It's in the cracks and crevices, but it's different.

It's something. I could take you over there and show you that Cresson Valley, you know, where they've dug that all out. I could show you the chimney of one of the volcanoes that came up there. It just cooled down. That's what caused all that gold over here to be in that area, was that volcano. When it erupted, it was enough that it pushed up all these other mountains around it. That's how they got here. Battle Mountain, over by the Vindicator, Little Bull, it's got all that gold in it. That's from that one volcano. It's something to look at. I go down through there with some of those guys in that pit, and one was showing me where some of the veins are.

I used to drive and give tours over there for the Mollie Kathleen. I drove tours through there one year. We'd take 'em down and go through the crusher and different things like that. This was one tour: take 'em down that way and go down to the ADR plant; show 'em how that all worked. The other was a surface tour that went around the top of it. Where the old Range View Road used to be, across the top of it. They had a place where you could look down in there, explain all that, go down through Victor. It was really interesting.

They're really touchy in there, and that's one of the reason's the Mollie quit takin' tours through there. We used to have two busses that we ran through there at different times. The insurance got so high, because you're going on mining property. The mine didn't really want to have to be puttin' up with us all the time. Because when you went in you had to check to make sure when they were blasting, you had to know all this stuff. Then you had to go and take classes, which I didn't mind, to drive in there. Because of the left hand driving there in that mine. You had to take different kinds of driving tests and that kind of stuff. We never really went on any of the haul roads; we went on the other side roads and things like that. It was maybe once in awhile you'd go down through there and takin' 'em down to the crusher you might go on a haul road.

But now that whole area's changed up there. Now I go up there to the American Eagle and look into that pit it scares me to death. It's really awesome to see that. You know, they get out more ore in one of them truckloads than we used to get out in probably six months. Maybe not that long, but they can haul about 300 tons in one truck load. That's an awful lot of rock. Course, they're mining in quantity. When we were minin' we followed the vein. Where ever those veins went, that's all the further we'd go with it. We was out to get as much of the high-grade as we could get, right outta that vein. Some of the veins wouldn't be any bigger than a pencil lead. Others would be bigger. You mined as close to that vein as you could. The veins run through these mountains for miles. It might be on a different mine, but it's the same veins.

Most of these mines are honeycombed together. One of the reasons was to get good air down in there. Another reason was one guy found a vein and he'd start following it to the end of his property, well, he'd tell the next guy and he'd start and go on through. There's a couple big veins they've been following for years from the surface, on the Bobtail. It goes all the way over to the Cresson, all the way up over there. We used to be able to walk from one mine to the other, to another mine, to another, ya know.

What about walking from Cripple Creek to Victor?

Well, you could.

Did you know anybody that did?

No. We tried one day on the Mollie Kathleen. Going over through the old Gold King, the Chicago, over toward the Cresson that way. But there was too many cave-ins. Used to be able to. We'd go down in on the Roosevelt Tunnel and come back up and go over to the Strong. Watson done that not very long ago. He come back over to the Strong to see that big slab of rock cut the shaft over there on the Strong. They wanted to go down in and see how bad it was from underneath, what it looked like from underneath. So they walked in the Roosevelt Tunnel. That Roosevelt was different than the Carlton. The Carlton was a lot lower.

What were you taught to call the frame that sits over a shaft?

Well, we used to call it the 'Gallas Frame.' It's the headframe, is what they call it. The headframe. When we were growin' up, it was always the gallas frame, because they had a lot of the guys would commit suicide, and they'd throw a rope over the beams on the headframe and hang themselves. That's where they got the word Gallas frames. During that strike, there was a lot of the miners where, if you weren't union or that, they took and threw rough ropes over them and they hung the non-union miners. There was a lot of that went on, not too much of it publicized.

I've heard you could tell who was running equipment by how they did it?

You mean the trammers? Not really. You had one certain level. Most of your trammers had just this one level to take care of. There might be another trammer workin' in a different area than you, but most the time you had one level, you had to pull all the chutes. You bring 'em out, drag 'em out to one track, then you go back with empties, load them up, bring them in on another track. Then the skippers would dump these cars. You just kept doin' that. Trammer was different than the skipper.

So the miner loaded the car and the trammer took it to the skipper?

Right. The trammer's the one's that loaded the cars outta the chutes. See, then they'd take it out on the Station. They'd have maybe four or five tracks. They'd pull in a line of cars here, cut off and take some empties back and load them, bring them out, put them on another track. When the Station got full, then they'd usually call for the skip. Then the skippers would come down and they'd empty that. Because they was usually outta the same chutes and things, so they went on top to the same place. So, you'd wait until you couldn't move on that Station before you'd get the skippers down there.

Did beaver swim up into the Ajax?

It could happen. I've heard of them building dams in the Carlton Tunnel. I know one time I walked outta there. Six miles of hell. Water'd wash holes between the ties, and then that heavy metals would hang onto the rails and the ties and make it real slick. You really couldn't see through it. You'd be walkin' in that water, you'd try to step on the ties, you'd slip off the ties, hit your shins. You could see the portal out there, walking closer and closer, but you're walking, walking, walking. I would never do it again.

Why were you walking?

Lightning hit the motor on the hoist; we all had to walk outta there. It was all the guys and everything. We took a tram first. They had these battery trams down there then. We hooked a bunch of cars on to 'em, the guys got into the cars and we run out as far as we could go until that battery went dead. From there we had to walk. That's the longest six miles I've ever seen in my life.

Does it get so mucky you couldn't drive the tram?

Yeah. The crust on those rails and everything. You know, if they went up and down a bunch of times and broke that off, it was all right. Once in a great while something like that would happen. The crust on there just slowed it down so much that you couldn't get through. I don't think they've even got track in there now. I think they tore all that track out, 'cause they was runnin' diesels. When John, my brother, was workin' for Texasgulf, I think they took the track outta there and was runnin' diesels.

Were there a lot of pranks underground?

They would do different things like nail their overalls. They'd be waitin' for the skip, they'd be asleep, and they'd nail their overalls to the bench. Little things like that. It was pretty serious most of the time underground.

What about Tommyknockers?

Well, they're true. There are Tommyknockers. You don't see 'em, but when you're workin' down there, you know they're there. And like the dirt that fell on me before I got buried, that was a Tommyknocker that warned me. Tryin' to get me outta there. Dumped dirt on me first, and I turned to get outta there, but I didn't get out far enough in time. That was Tommyknockers. They were helpful to the miners.

Did you hear them aside from that day?

Oh yeah, you bet. You can hear 'em ... lot of times sitting there waiting, you'd hear different sounds and things, you'd say, "Oh those Tommyknockers are workin' again." But they were in the mines.

Maybe if you hadn't gotten up you would have been hurt more?

Yeah, he warned me with that dirt. Just as soon as that started hitting me I turned and started to run. I got back far enough to where I was under the timber, and that timber, when it came, it slowed that slab down or it would have smashed me like a pancake. That timber slowed me down and saved my life, even though it did break, you know. But the Tommyknocker's the one that warned me, and I knew something was wrong. Boy, I turned around and started to run.

Did your dad believe in Tommyknockers?

Oh yeah. All miners believed. There was a lot of things that happened to miners, danger and things like that. They managed to get by and get out of it before it happened. Some of them didn't.

One of the worst accidents I ever seen in my life, workin' over there, I was trammin'. I had a partner pullin' chutes, and I was runnin' the tram. My partner was pullin' the chutes. He was taking a bar... the chute was that high up... he was takin' a bar and goin' up in the chute like that, tryin' to get it loose. That big rock come out, hit the end of that bar, and drove it right through the jugular vein. We had to... that was the worst. We had to get a cutting torch to cut that metal off and get him outta there. That was the worst accident that I seen on the mines. That was on the Ajax.

There was other accidents, not that I seen. I seen a guy got blown up, but ol' Doc Denman put him together, and a couple months later he was back workin'. His old helmet, his hardhat was just split. A rock come down and hit him, just tore that thing up. Thought he'd never live. Doc Denman was a great surgeon. He couldn't diagnose anything, but he was a great surgeon. He put a lot of these guys back together.

I was workin' on the Vindicator, I hooked up this saw. I didn't know too much about electric work. But anyway, I hooked it up to a 440 line. I didn't know 440. I went to start it, had little three prong switches on it you throw 'em in. I picked 'em up, threw it in, and that thing came back just burnt my hand somethin' terrible. I had my glove on, but it burned my glove off and everything else. It knocked me over the saw, clear up against the wall. It knocked me away from it, instead of hangin' on to me, or I would'a been dead. My whole hand was pretty burnt. And ol' Doc Denman he put tealeaves and this other stuff on there. Wrapped it all up, you know, he says, "It's gonna stink. Gonna smell like hell for a long time. But don't take it off." I went in to see him when he told me to, oh, about two or three weeks later. He unwrapped it; you couldn't tell it had been burned. He was pretty good about some things.

Colorado State Code of Mine Bells
J.W. Stonehouse Signs, Victor, Colorado ca.1900

You're 'Da Man

"Everybody likes you because their life is in your hands. Those miners would bribe ya t' pull all their muck out, so they can get on with their day."
~ L. R. Myers

Date of birth is April 13, 1953, and my first time underground was 1970. We went exploring the El Paso. Went into the Standard Tunnel with a couple of geologists, and over to the El Paso, via the Standard. There's cross-cuts and drifts, and a few chutes you have to go up to make it. Actually come out on the 2nd level of the El Paso, which is about 400 feet down. We found the tour level and wandered around. That was it. It was actively bein' operated as a tourist mine at that time, so didn't want to disturb anything that would infringe upon the tour that was being given the next day. These geologists, you know, they ended up breakin' a lot of rock, for sure. Just tourist rock. Little pyrite, little fluorite.

It was actually earlier that summer that was my first time underground. First I did tours. It was a summer job and I was still in high school. But yeah, the El Paso was the first hole that I ever went in, as the case was, and it was just to guide tours. Dale Weaver was operatin' it at that time.

What was it like to run tours there?

Oh, it was a little slower paced than what the rival Mollie Kathleen was doin'. So, it was excellent tours, even though we only toured one level. It was a circular tour, so you never competed with anybody else for your audience.

At that point in time the Mollie Kathleen toured the 700 foot level, back out to the Station and then down to the 10 hundred. On the 700 it was a dead end, so you'd have to pass the oncoming tours. It was very inconvenient because it was disruptive to the train of thought of the audience that you had. So it was. The El Paso was the best tour. Back in those days, an excellent tour. It was laid back. We had our fair share of customers and ended up with quite a volume of contacts. You know, people that I'd met on the tour, people from all over the world. Some cute, some not so cute. They all enjoyed the tour.

It was fun; it wasn't a lot of money. A certain amount of danger's involved with it. Babe Watkins was our hoistman. His days off, Dale Weaver'd run hoist. Babe was an excellent hoistman. He would tune out everything, but what he was doin'. Very, very thoroughly. Every move Babe did, he knew exactly what he was doin'.

On the other hand, Dale was a pretty excitable kind of guy running the double-drum hoist, and one morning he didn't counter-balance the small side. It wasn't adjusted for the depth he was lettin' us down. So, as he's lettin' the cage down, he pulled the little side clear up into the sheave wheel. Now that particular operation had a bell and scissor clamp on the little side of the skip, and the scissor clamp cut the cable loose. The dogs held the 'lil skip up in the gallas frame, and that inch and an eighth rope came peeling on off that gallas frame, hit the corner of the Gift Shop and tore bricks out of the corner of that building. We were very fortunate he didn't have a yard full of tourists, because it could have killed somebody.

Meanwhile, we're runnin' the level, down to the tour level there and the skip stops. So, we, Dave and myself, we ring him up on the emergency bell cord and he juggles us around a 'lil bit, and drops us down. That wasn't the command. And a little further, a little further. Finally he peels us clear down to the 1300. Bottom level. Dave and I wondered what the hell was goin' on, and finally he hoisted us on up and we realized Dale had made a major error. He was all panicky. Dave and I decided it wasn't that big of deal to get it set back up.

We clamped a rope to the cage rope, hoisted it up to the top of the gallas frame, come-alonged it down to the skip. Took log chains and chained that skip up to the timbers and re-attached it. Meanwhile, we're ready for business, we've got our only good pair of Levis on, and that rope-dope is worse than tar. We'd completely wrecked a set of clothes gettin' that skip back on line for him. He assured us, "Ah, no problem. We will take care of that. Oh, I'm so glad we can run, now we're back up." So, we got cleaned up as best we could, and we were back up. Open for business. Never did see any new Levis out of the ol' boy, but then again, when you had to beat each other to the bank to insure that you got your check cashed, you know, you can't expect him to buy you new Levis. That was quite an experience.

I worked there probably two summers and then the competition was pesterin' me about comin' up and runnin' tours at a "real gold mine." I had an old bumper sticker on the car: "El Paso Gold Mine Tours." George King used to give me so much grief about that bumper sticker that he decided in order for me to take it off the car, he'd have to hire me. I told him that the El Paso's a better tour, and he questioned, "Why is it?" Point being, it's a circular tour. And he took it to heart.

So I hired out for him in the spring of '74, and ran the hoist for him that summer, and that fall he retained Robb, and myself, and Cox. We drove the drift on the 10 hundred, down in the Silver Drift, and developed the circular tour that was used until Dennis altered it a little bit. A lot of the same drifts we drove are being utilized today. He's extended that tour since then. But it made just a one level tour at that point in time. So, ran the next year tour guidin' and runnin' hoist, and come fall then I went to work elsewhere. Then in '77, I hired out at Cripple Creek Gold Production's Mary Nevin Mine. With hoistin' experience, I hired out as Hoist Engineer.

How did you get that experience?

I put 150,000 people through each year at the Mollie Kathleen.

Did you need a special license to run hoist?

Oh, what it boiled down to was a physical from Doc Denman. Well, it's an official evaluation of health. That's it, and it takes a certain kind of individual, someone who can pay attention for minutes at a time, you know. The Mollie was runnin' 500 feet a minute, so in two minutes she's at the bottom level, and then you could get on with the poker game. (laughter)

But what's interesting is that hoist, at that time, had no safety equipment on it at all. No over-winds, under-winds, so you could peel out 1500 foot of rope, of course the cage would hit the bottom. That hoist could continue rolling until it picked itself back up, and never know it changed direction. No safety equipment at all. They do have safety equipment on it now. At the time that I was runnin' it, there was no safety equipment. I had nightmares.

Then I went in '77 to the Mary Nevin's and we set that hoist up. They brought this old piece in from Commerce City, I believe. It was a Hendrie & Bolthoff Hoist, about a 1910 vintage. But it was a good, old unit.

Was the gallas frame there?

When I got there, yes it was. The big orange one, yes. They were actually re-timbering the shaft when I got there, but they were using an air-tugger and just a snatch-block, and they weren't down very deep. So, we set the big hoist up and went to cleaning it out and re-timbering. The State comes in and inspects, and all the sudden we have to have all the safety equipment. Over-winds, dead-men. So, we complied. We put all this stuff on this old antique hoist to be compliant with Mine Safety Health Administration's standards. We continued to clean shaft out to 410 foot, and start development work. Ran hoist for 'em for about another four years then.

Up until this current mining project, it's the longest I ever worked for a minin' company. Usually you'd work the summer and a few months, and shut it down. So I worked a long tour there for them, then I went underground. I was shift-supervisor in the Twinem Stope. Conrad Twinem had a 'brilliant' idea of how we could go up and high-grade what the old timers left behind, and make some money. So in the Twinem

Stope, it became known as, I had a crew up there breakin' rock and droppin' it down to the 350 level out of the stope. The Twinem Stope was actually above the 250 level. See, and this is where Cripple Creek Gold Productions, in my opinion, had their mine plan screwed up. We should'a been sinkin' the Mary Nevin's main shaft deeper, and driving drift on down underneath the Henry Adney and the Lonaconing to get to the C K & N vein structure; down to virgin ground. Get right after the minin'. But there was a certain amount of ore left on the upper levels that was easy to get to, so they didn't make that the mine plan.

You were a supervisor but had done no mining?

Well, how hard is it? Everything's heavy. The lightest thing you grab is your loadin' pole. The next lightest thing is a box of dynamite, 50 pounds, and weight just goes on up to your drillin' machine. You don't have to have experience, to really know. You have to manage men. You surround yourself with men that know. Get the job done; encourage them. Get the job done. We were supplying a hundred ton mill. We refurbished that mill; put a new ball mill in it. It was a 6x6 Allis Chalmers ball mill, and had the capability of doin' a hundred ton a day. I trained a couple other hands to run hoist, and went on down where the action was.

What was it like to be the hoistman?

Oh, it was pretty noisy. Pretty good duty. Especially... what a pleasure it was, versus runnin' people out of the hole. I'd run the men to work every mornin' and usually pull 40, 50 trips up, pullin' rock, and then a few trips in between. Sometimes they'd come up for lunch, a lot of times they didn't. And then shift change. So it was really a pleasure to pull rock, because if you wrecked the conveyance in the shaft, odds were pretty good it was rock and not a cage full of people. It was very satisfying.

Was that a lot of responsibility?

Hoisting is, because you're relying on bell signals. You can't see underground, so you have to know what the miners want before you move that conveyance. The upside of it is, you're in control of that conveyance. You don't have to move it if you don't know what they want. Make them send the signal again. Or again, or again, or again. Until you know exactly what they want, and then you execute the command. Yeah, it was a pleasure working mine, versus a tourist mine. Oh yeah, it was a lot of fun.

The bells and the lights are what you have to pay attention to?

Well, that and the boss talkin' to ya. Which is against the law. Nobody is to speak to the hoistman while the hoist is movin'. That's law. Yep. So yeah, you focus on what you're doin'. You tune out everybody else, and everything, and do your job. You tune the world out when you're in operation, and when you're not in operation you tune that out.

Did you hear bells in your dreams?

Oh, at the Mollie, that was a nightmare. I rescued people out of that thing more than once, in my dreams, my nightmares. No safety equipment on it. Oh, it wasn't good. It was a tough way to make $2 an hour. It might'a been a little bit more. That was good money back then.

Was hoist a lonely job?

Oh, no. You're 'Da Man'. Everybody likes ya. Everybody likes you because their life is in your hands. I mean those miners would bring ya up high-grade specimens and bribe ya t' pull all their muck out, so they can get on with their day, you know. We had a contractor down on the 400 level and he was contracting, that's where he made his money was to get footage. To make footage you have to get a round out and get a round in, and you can't be getting a round out if you're waiting on a damned skip to get back down to ya. So, that's where the bribes come on in.

The bribes worked?

Oh yeah. That changes everything. "Heck yeah, let's get the old man mucked out right away." Ignore everybody else, unless it was convenient to drop somethin' off to 'em. I was goin' anyway. But you focus on moving material. Production. That's how the small mines make it, is you need production. You need a dedicated hoistman that can sort that on out, and realize to keep that mill runnin' you gotta get that ore to it.

I've heard you could tell the hoistmen apart by the way they ran the conveyance?

Oh, it's like any piece of equipment. Yes, yes. You have good hoistmen, you have better hoistmen. How smooth they come into the station. How accurate their stops are, which is critical when you're dumpin' a tram car into the skip. You want to leave, oh, about a foot above the level. That way the trammer, that's trammin' that muck out to the skip, can run it right in. He hits the side of the skip. It's high enough, he doesn't loose the car into it. It can't be too high, or he can't dump the car into it. So, if it's spot-on, then it's efficient. Dump that car right into the skip, ring him off the level. The skip gets to come on up, hit the scrolls and dump it across the grizzly. That was sweet. Production.

I think to be a'people hoistin' you have to exercise the same degree of accuracy. Because you don't want 'em to trip. You can't overshoot the level, or you can't get the cage door open. If you're too high, it's a big step. So, it's a matter of precision. You have to be a damn good hoistman.

How do you know where the conveyance is?

You have a dial indicator that's indicating the proximity of the cage. As the ropes peel off of the drum on the hoist, there's a chain-drive, usually chain, drive over to the dial indicator that's calibrated to where the level is. So when you get close to the level, you have spots on the drum, see, you start watchin'. And then like, your indicator you know, when it's "X" amount above the level, you're 3 wraps off the level, well, you're slowin' down. Then you start watchin', countin' the wraps comin' off the drum, and then you ease it on to the spot that you have established on the drum, and you hit it, every time. Give or take a 'lil bit.

Were you a good hoistman?

I was a damned good hoistman. Damned good. One of the best. (chuckle) The only person I almost dumped in the sump, at the tourist mine, was Johnson. Night watchman was havin' a bad day and he was pickin' on me, and I was peeling rope off at 500 feet a minute. He was in my face, I was in his face. He was hung-over pretty bad, and I wasn't takin' it that day. I spun on around, and I looked, and I was one wrap off of the ten hundred, goin' down the hole wide open. 500 foot a minute. Whoa. Those brakes came on; come to a screeechin' halt. Yo-yo'd him in the cage, in the shaft there. Which is good, you want stretch in the rope. When that rope stops stretching is when it's fatigued and about to break.

Did it bounce people around in there when it would yo-yo?

It had a tendency to slam them around. That's indicative of a poor hoistman. You know, poor hoistmen would do that. Well, locked up on both drum brakes, looked down and sure enough he was spot-on the station. Had an occasion to come clean with Ted and told him he was the one I almost run into the sump. Which wouldn't have been bad, but the sump was only like 15 feet underneath the bottom level, so he would'a hit bottom pretty hard. But I come clean with him and he says, "I remember that day, I remember it well. I thought you were comin' into the station pretty fast, and every time I'd start opening that cage gate, I'd yo-yo past it and back up, and it took a little bit to settle on down before I could get the gate open." That was almost a disaster. Just the way the incident was structured up with that pump. But having very fast response time, and good brakes, I was able to shut it on down and get him parked. It was scary.

Were there other times you were scared running the hoist?

Well, there was one other time, that I had a tour guide forget to close the cage gate. He had a cage full of college ladies, and his mind wasn't on his job, and he didn't lock the cage gate. The conveyance sat on the collar, that's on the surface there, for thirty minutes or so, and then had a call on ten hundred. So, I dropped her on down the hole and it hung up in the shaft. It found a level that wasn't on the indicator and the rope went slack. Well, the night watchman was on the surface, so he crawled down the manway. He rang me up and down, and he ran the shaft the full length, then came back up. It was quite a wreck; it bent that conveyance up pretty bad and the cage gate wouldn't lock anymore. The door had opened up above the 200, right around the 200 foot level. It opened up and caught the level, and it tweaked it.

Well, we had probably 12, 14 people in the parking lot waitin' to go down. Had tours underground. We went out with jacks and hammers and pry-bars. Went to overhaulin' that cage and watchin' them people. It was quite a sight, because they just kept buyin' tickets and watchin' us fix that wreck. We kind of joked amongst ourselves that if we came on a wreck like that, we'd be in wantin' a refund immediately. Immediately. No way in hell we'd be goin' down the hole. But tourists bein' ignorant, they continued to buy tickets. We were down an hour and a half. George King drug the tour guide on out back and commenced to chew his ass. What he was chewin' him about was, "Do you know how much money you're costin' me today?" Not the fact he had people down the hole.

It was an extended tour for 'em; they were down probably an hour and a half. Tips were probably pretty good for those tour guides. That was the only other close call incident that ever happened.

That's really lucky?

Oh, that's bein' good. That isn't luck. That's takin' care of business.

Is the shaft one of the more dangerous places in the mine?

Well it could be, yes. Everything falls down the shaft. You're talkin' to a man that never wanted to be a shaft-hand. Everything comes down the shaft. Little rocks, big rocks, a drill steel or tram cars. All that stuff comes down, and in the main shaft there's no place to get out from under it. Now, in the conveyance there's a bonnet over ya. There's steel protection over ya on that.

What if something shifts in transit?

Well, securing the load, especially suspended loads, is very critical. You really make sure your load is secure. Drill steel can get away from ya. Get hung up and come out of the cage. So anytime you're transporting tools or equipment, you're usually runnin' slower than what you are if you're pullin' rock. Especially if it's over-size or a spooky load, you just take it real easy.

Was the ore at the Mary Nevin good?

Oh, it was good enough that the top-landers at night would take all the lights and put in the ore bin, and high-grade the ore bin. Run all the mine lamps dead. Go down the hole the next morning and your light would be dead in an hour. Oh, yeah. It was part of the job, so they thought. It was the advantage.

People seem to feel entitled to some of what they mine?

Yes, they do. Well, the thing about it, too, is upper management decided that dog-gone-it, the miner's are entitled to a specimen of the rock they mined. I am surprised they didn't run flood lights out. (laughter)

What would you say about hoisting?

It's a livelihood. Hoistman is a great responsibility, but it's a good livin'. If you don't mind the responsibility, it's pretty good duty. You're in the hoisthouse; you've got time to polish up the hoist, 'Armorall' the motor, and 'Turtle Wax' it. Took care of your equipment. It's work. It might not be real physical, it's more mental work. It's not manual work like the miner's are doin' down the hole, but it's mental work, and it's tiring. It's tiring and it'll take it's toll, because you have to focus. You have to pay attention. But then upper management got to the point they thought, well, you've got all this time on your hands, you can overhaul the drilling machines, and all this other stuff. Well, production picked up. Didn't have the time to do all that. (chuckle)

Did the hoist ever break?

Well, yeah. You have to replace the wire rope. We did break a bearing block on the main jack-shaft. And of course, you have to keep it working, so used the local welder to weld it up, and then that got us through until we could have some new bearing blocks machined. They were old Babbitt bearings. Not really lead, but it's like lead, though. We had to get new bearing blocks machined, and new Babbitt poured in it.

It was a lot of fun, and a good... you know, these smaller mines, you really get the sense it's a family affair; everybody's in it together. We had Johnny-come-latelys come in, and we had tramp miners come in. But, it was good duty. That little tour lasted, oh, I think seven years, anyway.

Were you there when the Mary Nevin closed down?

No, I'd moved on. I went to work for Silver State Mining Company, bein' a chemical miner. Mr. Hunter, Ed Hunter, came down and toured the operations, being a mining man, and I think he was impressed with how clean my hoisthouse was, how clean my hoist was. At that point in time, Silver State had been runnin' about a year, and they was having problems with operators tearin' up equipment. He suggested I come up for an interview and run some equipment for them. Because I took care of my equipment.

I ran a 980 wheel loader. I was a mill loader. I loaded vats, unloaded vats. Loaded the crusher. Oh, good duty. Great big smile on my face. In three passes with that loader, which was less than ten minutes, I could move the same amount of material it'd take me a half a shift to do underground. I mean that brought a smile to my face. Progress. Now we're mining. Production. It's all about tons moved, gallons moved. They leached in 1,000 ton indoor vats. That was a pretty sweet little set-up. I worked for them until they sold out to NERCO.

I lived in Cripple Creek, you know. You work wherever there's work. So I come back to work with NERCO, as loader operator. They went into this fancy phase, built 2,000 ton vats. Woo-woo, let's go big time, huh? So when that was done, I went ahead and moved into Process. I was upgraded to Process Supervisor. Worked in the control room, and just moved solutions. Pump water.

Not quite like being underground anymore, were you glad for the change?

Oh, at that point in time it looked like that's the direction mining's goin'. If you're going to stay in the mining game, you have to change with the times and the technology.

Ajax Mine

Well, It Can't Go That Far

"It was a big truck tire, and that's a very steep dump up there."
~ Jack Lorenz

I was born on October the 8th, 1935. There was mines all over here when we were just kids, you know. They had, I think it was six grocery stores, five bars, beer joints, and they were all going great guns. There was a lot of miners around spending their money. The mines were going pretty good up until when the war started.

I worked in a machine shop for my dad at the Ajax Mine, up on top of the hill. The last mine running was the Ajax. They shut it down in 1962. January the 2nd of 1962. It was the only mine runnin' then. The only place to work, really. They shut it down, and then I moved away and went to work at different things.

What kinds of things did you do at the Ajax?

Worked in the machine shop. Welding, machinist. Oh, I don't know what you'd call it, tool and die, electrician. Did it all. They didn't have any Unions, or anything like that. Sometimes I went down in the mines and worked on equipment down there; not very much.

I walked through the Carlton Tunnel about three or four times a year, just to see that it was clear. The seven mile tunnel, drainage tunnel, that goes into the Ajax mine. That used to be my job, to walk through that. If there were beaver ponds, or muskrat ponds, or stuff like that, we'd break them. They'd live in that tunnel. Warm water and these beavers and muskrats would get up inside of the tunnel. They'd dam it up and then they'd keep damming it up, and pretty soon you don't have drainage. The water would back up into the mine.

People say the water's bad, but there's fish in it. There's frogs in it. It's kind of a yellow looking muddy thing on the side of the road. People think, 'Oh, that's bad.' But it's just the minerals coming out of the mine. How could these animals live in it, if it's so bad? They'd go clear up into the mine where they were mining. They'd catch 'em in there sometimes and bring 'em up on top. The beavers. Take 'em up and put 'em in a box and bring 'em up the shaft, and turn 'em loose somewhere. There'd be a lot of snakes in the tunnel. Usually right at the portal where the water is so warm. That's the place to be when it's cold outside. You've got nice warm water in that tunnel coming out year round, twenty-four hours a day.

I don't know now, they say it's pretty well plugged up now. Nobody's taken care of anything, and that stuff gets caked up on the rails and starts making it's own pond. The mine don't maintain it. They cleared it all out a couple of times, went clear through it, you know. Dug it all out, you know, they had little ditch witches and stuff like that. They cleaned it all out once.

It's just drilled rock almost all the way through. It gets plugged up with all that stuff you see on the outside, that yellow looking stuff. When the mines were running good, they used to run those big battery motors through there. Great big, heavy, piece of equipment, and they'd kind of crush the track and clean it up. You could drive through it then, you know, with the battery motors. They had to do that every once in a while, or it'd get that way. Then they'd have to clean it all off to get through. They claim it's about 7 miles. You can see, it looks like a dime when you first start walking, then you walk forever to get out. You can see it, that's how straight it is.

Did you once roll a tire down the Ajax dump?

Uh huh, I did that. Well, it was a bad, bad, deal, and I'd never do it again, because it kinda really spooked me when I did it. It was a big truck tire, and that's a very steep dump up there. And I was up there, the reason I was up there was probably looking for some kind of scrap to make something out of, you know, we had to machine everything we'd need.

This guy that was with me, Earl, I said, "Help me get this tire off, I'm gonna roll it off this hill." He said, "Oh no, I'm not helping you. I don't want nothin' to do with this, you could kill half the people in town." I said, "Well, it can't go that far." You know, just down below, there's kind of a pocket down there, and that tire couldn't get out of there.

I got it up myself, and I inched it over there, and I debated quite awhile if I was going to kick it off. And I finally did. That thing got about half-way down towards Victor and turned and went over by Squaw Mountain, where the "VHS" used to be on the side hill. It started over on that mountain, I thought, well, that's fine, it isn't going to hit Victor now. It got half-way over on that hill and it started wobbling, and it turned, and away it went toward Victor. It went through somebody's chicken yard, I could see that. It went through fences and a couple of old shacks and cabins there, it mowed them down. Then it finally hit a dump and stopped, but it's a wonder that thing didn't go through the whole town. Didn't do that anymore. Enough of that.

El Paso Mine

Mining is a Good Way to Go in the Hole

"This mill, all of my hopes, all of my dreams, all of my efforts... just went up in smoke. I just almost felt like giving up. Giving up living." ~ Dale Weaver

I was born in June, 1918, at home on a wheat and dairy farm in Kansas. I was one of a family of 16 children. I was down the line a ways, towards the end, and my father was a preacher in a little country church. I went to school through the 8th grade, then with delays I went through high school. It wasn't all at one time. Took me awhile to get through high school.

I was working on the farm all the while. Dad told me, "Can't let you go to high school, you're the only boy left, and I gotta have you here on the farm, so can't let you go to high school." So, a couple years later, he let me go to high school. It was just life. He did the best he knew, I had to conform.

It was quite a few years later that I got here. Can't remember when I came here; think it was '62. The first time I went underground was in the El Paso, after I bought it. I bought it before I even saw the mine. Had the Superintendent take me down some days after I'd bought it. Said, "I want to see what I bought." So he took me down, gave me a tour. Explained to me the most important features, the most important information about the mine. That was my introduction to underground. I loved it. An adventure. Had no apprehensions underground whatsoever, just seeing something different. Something new that I'd never seen before. And especially since it was mine. It was very impressive to see that.

Did you run mine tours there?

For three years. Had good tours. Most people liked it. They had only two tours in the District. The Mollie Kathleen and the El Paso Tour. There was a lady over in Victor, a tourist asked her one time, "Well, I hear there's two tours here. Which one's the best?" She said, "Well, you want to see a good, well-organized, sensible, informative tour, take the El Paso. You want to see a circus, go to the Mollie." At that time, it was a pretty good description of it. It was just hype and trying for effect, and trying to inflate everything over at the Mollie. "You want to see a circus, go to the Mollie."

I hired people to be guides. I was running it; I had to hire guides. John Fuller, Pete Pope. Oh, I had a number of local people that were guiding. Dave Reid, he was a guide for a while. Oh, my mind is not so fertile as it used to be. We called him Robin. Everyone called him Robin. He's workin' for the mill, has been for all these years. Robin's his nickname from a kid; we called him that most the time. He was a good guy. He and Dave worked at the same time and they were always up to something. Always getting something different on somebody all the time.

Pete Pope used to live in Victor. He was a guy built like a barrel. Short, wide. Wasn't really fat, he just had a lot of stuff to him even though he was short. He was not anyone to put on any airs, to try to inflate himself or anything. He'd just talk. Just blunt. No wasted words and people loved him. They knew that he was an honest man. He'd tell them what he knew, and he knew what he knew very well. And they just loved it. Not trying to put on any airs a'tall. Chewed tobacco all the time. He'd be standing talking to a tourist; he'd just spit, right down at their feet. People were shocked at it. Usually they'd just begin to smile. (laughter)

You thought to mine at the El Paso?

It had been idle for years. It had tours about four years before I bought it, but no mining done in quite a few years. So I just went to starting. I had tours there, and had to get the rest of the mine in shape. I formed a small corporation. I was the President and the Manager. It was a number of years before I opened the mine. For a while I was just by myself. When I started work, I had to have some employees. There was no mill here to process the ore, so I had to build a mill. So I did that; I managed that. Finally got the mill going. Can't remember exactly what year it was, but it was in the '60s.

How did you go about restoration?

Well, you bring in timbers. You measure your face where you have to put the timbers. Have a man on top to cut the timbers according to your measurements. Then they sent the timbers down and you just installed the timbers, and re-timbered it. New framework and everything. It's a lot of work. There were a lot of difficulties in building it. I didn't know much about mills. In fact, I'd never seen a mill run, before my own did.

I had never been in mining before, so I started from scratch. Had to learn as I went along. Did a lot of reading, and talked with people that were experienced. Scoured the whole state of Colorado, Arizona, and a couple other states to get the equipment. Took it up here and there. Got it here and finally got it installed. Had the thing about finished, was just checking it, getting the bugs out of it. Finally got it to where I was about ready to hire three crews, to mine and mill 24 hours a day. Somebody came and burnt the mill down.

I went to my brother's funeral in Iowa, and I was gone for several days. Came back, late at night, and stopped at the post office, which was over here at that time. Larry drove up and said, "Hey, you know there's a fire at the El Paso?" I said, "No, I just got to town, hadn't heard anything." He said, "There's a big fire over there."

So I went over. There was the mill, just in the last stages of burning down. All that mill, I put in to... all of my effort, all of my knowledge, all of the knowledge I could accumulate from others. It was pretty well finished; just about ready to start into mining and milling, 'til somebody came and burned it down. Well, that was almost unbelievable. Almost unbelievable situation. I had been so busy, taking care of details and stuff, and just getting along. It was the end of a big dream. I couldn't recover. Didn't have enough money to recover. I wasn't able to do anything; I had no resources. I didn't have money left to keep going; I lost the El Paso after the fire.

And then, I think it was the second night, I just realized what it meant. Realized how significant it was. This mill, all of my hopes, all of my dreams, all of my efforts just went up in smoke. I just almost felt like giving up. Giving up living. I couldn't come to any kind of solution. It was just pure disaster. I couldn't sleep, just thinking about how serious it was, and what the future would be. All of a sudden a little tape recorder began playing in my head. Or it seemed like. It was very simple, the message was. It just kept repeating, like a tape recorder on repeat, all the time. It just kept going. And this is what it was:

(Singing melodiously) "God is so good, God is so good, God is so good, He's so good to me." It just went over and over, all night long. I think that's the only thing that kept me from going off the deep end. I had something positive in my head, after all of that incredible, terrible, awful, significant news of the mill burning down. The mill no longer there. It was like I had no future at all. That got me through it and I was able to get rational, and see things in reality instead of just from fear and all of the negative stuff. That was God. He planted that little tape recorder in my head, and he kept the joy as long as needed. That was a big part of the whole thing. Sounds simple, but it was very significant.

I just lived. Wasn't much I could do. Didn't have any money to rebuild the mill. Had to forget that. I just went on with my life, with whatever else I could.

Do you think it was burned because somebody didn't want you to succeed?

I figured it out. It was jealousy and frustration, because he wasn't getting anywhere in his plan. He was trying to be a big shot. He thought mining would be the fastest way to be a big shot. But he couldn't do it. He tried three different companies to get into mining; they all failed. Finally, out of just sheer frustration, he had to get even with the world. So he burnt the mill down. He couldn't have it, why then, I couldn't either. The whole thing just fell apart. I wasn't able to do anything because I had no money left. No help. I just had to back off, forget about that.

I worked for some other mines, and did other things. Didn't have a definite plan. Didn't have anything lined out. Just had to keep plodding my way along. There was a mining company attempted to start up here in the Victor area. I worked for them for a while. Then I worked for Southway. He was a contractor that came up here. I worked for him a short time, 'til I couldn't get along with him.

I just gradually had to adjust to it, and after a while, instead of being angry, violently angry, at the fellas that burnt the thing down, I just decided, well, I don't have a mine. But I'd rather be me than the fella that burnt the thing down.

They saw the smoke in Cañon City and they saw the light from the flames down there, too. It was at night, so they saw flames. Saw the glow from all this distance.

Did it hurt the gallas frame?

Yeah, it did. It weakened it. Never has been used since then. The fire was the end of the life of the El Paso. It has a future, but it'd be very, very difficult to get it going again, because the shaft is caved. After the fire it caved in. I was working on it at the time. It caved down; the first 90 feet had caved. And I had started that upper part. Then later on it caved on down from there, I don't know how far. You can't see it. It caved right on down for quite some distance. Probably 700 feet, that's about the way it is now.

I've heard that you have at least 9 lives?

Well, I was crushing ore, this was years later. I had the mine where I live now, the Hiawatha Mine. I was crushing ore and leaching it myself. A big rock was in the feeder, and it stopped the flow of ore coming down into the crusher. So instead of shutting it down, which would have been the smart thing to do, I just crawled up on top of the crusher, where it was running, and something happened that I can't quite figure out.

I think the most likely thing was the that wind had blown my pant leg and caught under the belt that was going around, and it caught me and just pitched me head first into the steel framework of that feeder. It was pushing against my rear end and holding me down. I couldn't get out. After that I guess I twisted in the right way, freed myself. I stood up and I couldn't hardly see because blood was running down my face so fast, I couldn't see anything. My scalp was peeled completely off my head, and was hanging down my neck.

Almost immediately, I felt I was passing out. I quickly crawled off that crusher and laid down on the ground, on a slope with my head downhill. I did that deliberately, so the blood would go to my head better, so I could recover from blacking out. When I came to, I couldn't get up and walk. I had to crawl on hands and knees to the tractor that ran the crusher. I pulled myself and shut off the gas, so it'd kill the engine. Then I walked over to the generator and shut it off. I'd got on my feet by that time. And then I walked over to my car and drove myself to the Clinic. I remember clearly, it was such a pleasant morning, such a beautiful morning when I was driving in to the Clinic. I drove in, drove up to the emergency entrance, laid on the horn. Lady came to the door, stepped out and immediately went back in. I tried to holler at her and say I need help. But she knew what she was doing; she got the EMTs. They came over, and I had done all those things, I had gotten up, I had shut off the tractor, the generator, and drove in to town, no trouble. I got here, but I couldn't get out of the car. Perfect illustration of God's powers. He will provide for all your needs. God knew what I needed, so he took care of it

So they got the helicopter and flew me to the hospital. That was a dirty jip. I like to fly, and I like to look down and see the scenery. They wouldn't let me up, had me lying down all the way there. Couldn't see anything. Then they were sewing me up. The doctor sewing me up said, "You know, there are three layers I had to sew up." You got three layers on top of your skull; he sewed all three of them. He sewed me up and put me in a CAT scan. And they were perplexed, puzzled. After awhile they decided to put me through the CAT scan again, because the way I was torn up... thigh meat was torn up to the bone, meat was torn up on my arm, and the scalp was completely off my head. They couldn't understand how someone who was torn up so badly didn't have any fractures. So they put me through again, still no fractures. Finally had to admit, there were no fractures. Actually there was one fracture; this bone in my little finger was cracked.

So in the evening the doctor said, "You're ready to go home." Without thinking at all, I knew that I wasn't going to go home. Some very unusual thing, but I knew without anyone telling me. About a half hour later he came to me, relaxing on a gurney, he came over to me and said, "I changed my mind about your going home. I sent samples to the laboratory, and when you have a lot of torn tissues your body gives off a kind of protein that it doesn't give off any other time. It's in your blood, in your urine. I've never seen results that high. So it'd be better if you were here, just in case." I said, "Well, I knew I wasn't going to go home. I just knew." "God must have told you." I said "Yep." Yep.

Next morning I called some friends, good friends. She came in and got me, took me to their home. The next morning, getting ready to go to church, I said I want to go to church. They said you don't have any clothes fit to go to church. Well, she went and borrowed from a neighbor, and I went to church. I stayed there until Saturday, then went home and went to work again, crushing rock. With gratitude. That was one thing that happened. It was the worst. But that's a perfect illustration of what God is like. You can do anything you want, but not without God.

At the Hiawatha, how did you get ore?

I was processing dump ore. Leaching. Small scale cyanide leaching, like the big boys do. I did it on a very small scale. Bought some semi-trailers, would load them with ore, I'd crush it first, then put the sprinkles on top, in that semi-trailer. The cyanide solution collected. It saturates that ore in the trailer, trickles down and runs out the bottom. Put that through a charcoal collector, the charcoal collects the gold out of the solution. Then I would be able to retrieve that gold out of the charcoal. I learned a lot from other people that knew more than I did.

Can pine oil be used in any way?

Flotation. I used that at the El Paso, before the fire. I was running ore, mostly on a test basis; get the bugs out of the machinery. I did process some before the fire. I was doing the flotation process.

You're quite an adventurer?

Well, sometimes you have to be. If you're going to make any progress, there are things you have to do. It depends on what your incentive is. How urgent it is to you. People do the things that are most urgent to them. The things that have to be done. I learned a whole lot more just by doing it, than by studying it.

What do you think about Tommyknockers?

Oh, it's an intriguing little folklore, is all it is. Folk story. But if people believe it, then it's true for them.

What about the air underground?

No problem. Some mines had bad air. The El Paso was very well vented. See, it's connected with the big tunnel that runs from down in the valley, clear off to north of Victor. So there was circulation there all the time. The Hiawatha mine had no problem. Only a small percentage of them had bad air.

I got the Hiawatha later, about 1968 and have lived there since then on. I didn't do any work in the Hiawatha. I went down, but I didn't do any mining there. It wasn't in shape to mine. The ladders were bad. Had a hoist, but there wasn't anyone there to run the hoist when I wanted to go down. So I went down the ladders. That was none too good. First time I went down, the ladder collapsed on me three times. I knew they were bad, I took precautions. Put a rope around my waist, and then I put it through an iron loop up there on the headframe, and held the other end. So when the ladder collapsed I just gripped the rope and let down easy. You don't trust luck, because luck isn't trustworthy. It's very shallow. Only 300 feet deep. It's a baby mine. El Paso is 1300 feet. The Ajax mine over there, that's in Victor, 3300 feet.

Do you think it takes a certain person to like the underground?

I don't know how you could characterize such a person. Some do, some don't. It's personal taste. Some people let fear dominate all of their decisions. But for the man that realizes that you can't live on fear, you have to live on reality, you've got to live on the likelihood of things, those people will go underground and like it. You take a kid and tell him all of his life, don't do this you'll get hurt, or don't do that you'll break something, all those cautions like don't go out in the garden, there might be a snake out there. There was one seen there three years ago, there might be another one there. All these fears the mothers can hardly resist telling their kids fifteen times a day. Don't do that you might get hurt, don't do that you might get lost. We teach the kids to be timid. We teach them to be scardey-cats. Most mothers do.

Is there anything you would add about mining?

The sum of my experience in mining is tied up in this statement: Mining is a good way to go in the hole.

International Mine Conveyance

Commander Drum, Beaver & Young Men End the Era

"I liked that underground life. The work was hard, but it was neat because not very many people had the foggiest idea what was goin' on down there."
~ Brian Hayes

We came to Victor and literally had $20 bucks in my pocket. That was it, and two kids. The first job came along, I had to take it. I was just glad to be out here in the mountains, that's all I thought about. It was winter. We came in October of 1973, and I think a month later I got a phone call, or talked to somebody, and they said "Hey, do you want to work on a drill rig?" and I said, "Sure." You obviously had to make money.

Two bucks an hour. Two dollars an hour, working up at what is basically the very first, the very start of this big open pit now. We were drilling up in the top. You remember the old town of Stratton? Drilling up there, and that laid the groundwork for everything that came afterwards. Twelve hour shifts; noon to midnight, and midnight to noon. I think it snowed every day. I worked there through May, and the guy, I forget what his name was, but he would come around. The name of the company was AZCON. It snowed every day up there, all that May, and I think I got another job working roofing, and building and so on in June. And just the same day I was going to tell him I found something else, the guy came around and said, "I'm gonna give all you guys a fifty-cent raise. You'll be making two-fifty an hour." I was just glad I had somethin' to say, like "No."

Then I did construction. Roofing, mainly. That and odd jobs, and so on. Whatever we could do. Then somebody once said... I don't know how this came about, if you remember a guy named Dick Jackson. He had been seriously wounded in Vietnam, and was kinda getting his life back together. He ended up coming up here from Texas, I don't know why, or how, but he came up here. He'd lost an eye. Heck of a nice guy, he was at kind of loose ends at the time. When he was like 17 years old he lied about his age and was in Vietnam at 17. On the second tour of duty, which would have made him 18, he was seriously wounded and had lost an eye. Anyway, he said, "Hey, they're hiring miners in Leadville. Let's go up there." So he and I and a couple other guys all went up to Leadville and signed up. Ironically, of course, he only having one eye they wouldn't hire him. But they hired the rest of us, at the Climax Mine.

That was my first time underground. It's a huge mine. When we'd get on the mantrip in the morning there'd be, you know, you've been up Freemont Pass. You'd go into the dry, the change room, and then that was also connected to the portal of the tunnel. You'd go directly from the change room right into a mantrip. A big train would carry everybody, say at least a couple hundred of us, you know, on a shift, we'd all get on that mantrip and go around.

It was a train. You could sit two on a seat, and there was an opposite seat. You had this thing over your head, you know, for protection, big heavy steel. There's a big string of these cars and each one held four miners, two of ya facing each other. That train would go around and drop everybody off at the different work stations, and come pick you up at the end of the shift. They also had another level of the mine that went down deeper, some of the miners worked down deeper.

What did you think the first time you went in the dry?

I thought holy buckets. I immediately liked the other miners. They were just really nice guys. There were some women that worked there, too. It seemed to be kind of, seemed to get along real well, and were helpful, and worked hard. But it was a very extensive mine, you could get lost.

I remember working for a guy who was on what was called the repair crew. Went around and fix the track and tear down the cement forms and a bunch of stuff like that. This guy said, "Hey, go back to the shop," we had an underground shop for our crew, "go get some such and such a thing." I said, "Alright." We were up in a stope and I climbed back down and I started walking, and I got lost down there. I walked and walked. All the cross-cuts in there were numbered, and there was all kinds of graffiti sprayed on the walls in there, so, I said well, I've been by here because I remember this piece of graffiti.

But I finally did find it and when I came back he wondered what the hell was taking so long, because it wasn't very far at all, really. I just went down through the stope and headed off the opposite direction than I should have. Anyway, Timmy was a year old. Rosemary came up there one time and said, you know, the winter was starting and it'd be cold and we lived in that little house over here, that was a shack. She said, "I think you'd better come back to Victor." So, I quit the job, didn't even go for that shift. And came back here and then I started working at the Gold Bond shortly thereafter, with Al Goodwin. Yep, and the Pratt boys. Because I had known the Pratt boys, John, Larry and Greg. They all worked for Al over here, and a guy named Charlie Nutall. I had known them from bein' around here for a couple years, and one of them, Larry, said, "Hey, you wanna come over to the Gold Bond? We need a top hand." So, I went over there and worked there for a couple years.

If you come outta Cripple Creek and you go past Ma Beards place, it sits down in the little Valley, then the road comes like this and it heads down to Anaconda. Right after those "S" curves it goes down. It's on the right hand side and it goes down and at the bottom is the Gold Bond Mine.

It was 400 feet underground. They had one level that was stoped to the surface. You could get off at the 100 level, where we made our primer, you know, dynamite primers, because you had to have that separate from where you were working, obviously, and you could see all the way up to the surface. That was at the hundred, hundred and twenty-five foot mark. Then we were down in the 400 working for Al. We went through some different people. Mostly Pratt brothers, and then Brian Marshall and I worked there together as partners for quite a while.

At the same time, of course, gold was going up. In 1975 gold was on the free market, so there were a lot of speculators around.

Did that really change things here?

Yes, there were more jobs. By this time I liked that underground life, the work was hard, but it was neat underground, because not too many people knew about it, you know. That was one of the real neat things about underground hardrock mining. Not very many people had the foggiest idea what was goin' on down there.

What did you think the first time underground?

Never really hit me too much until I worked in a deep mine like the Ajax. It was a different world completely, full of water and gas and heat. At the Gold Bond you just never you really thought of it, it was just 400 feet deep and everything seemed pretty safe and normal.

The hoist was a Ford V-8 motor that they had hooked up to a drum, and that's what cranked it. It was like truck motor, ran off of that. It was a real slow moving little one-ton skip.

What did you do there?

Al had the idea that, you know it was owned by a family who are still around here, they're from Texas. They still actually own that gold mine, but anyway, you know Al, he was kind of a promoter guy, and he had talked them into this idea that the mine across the little valley there was the Caledonia mine, it only went down a couple hundred feet but had a pretty good ore body, proven ore body, that he could drive. They must have owned some rights on Caledonia, too... that he could drive the Gold Bond right underneath this ore body and be rich. There would be a big deposit of gold. He'd keep changing his mind where it was, so that tunnel that we drilled, that cross-cut and drift that we drilled down there is just, it meanders all over and it probably goes about back for a thousand feet. Never did find where it was.

Did you get any ore out?

No. Not while I worked there. See, I worked there for a couple 'a years, with a couple 'a lay-offs in there. Because, you know how that goes, no money. But at that same time, as I said, the price of gold was goin' up and the Mary Nevin started up. So while I was working at the Gold Bond the Mary Nevin was goin'. The Mary Nevin was like $7.50 an hour. Oh yeah, let's go over there. At the time that was still low wages, but I think Al was paying us like six bucks an hour or something. So, I think in one of those lay-offs, I don't think I quit at Al's, but I think it was at a lay-off I went over and talked to... because a couple of the other guys had gone over to the Mary Nevin. Charlie and all the guys around here; that was the big employer. They employed like 15 miners at the Mary Nevin, spread out over three shifts and they had a little mill operation. You remember the Mary Nevin days, don't you? You'd have been in grade school then. 1975-1980 or so. It ran, with a few lay-offs in there, all that time. I went over there and worked. Then at the same time, '77, '78, somewhere in there, the Ajax, they were gonna start doing some work.

It had been running, but with just a maintenance crew. They started hiring, then of course people from the Mary Nevin, because all the sudden, for the first time, instead of these small poor-boy operations, this was a big company. Texasgulf. It was like oh, this big mining company's coming in here, the money's gonna be better, and they're gonna be here forever, and they're gonna strike it rich, and all that kind of talk. And they had a benefit program. Even helped people buy houses here in town, remember that? There was quite a few, like Roy Pulley, he bought a house. Give 'em like, I think it was $8,000 and no interest. You had to pay it back, but no interest. They wanted them to live in town. Quite a few of them bought houses.

The Mary Nevin kind of struggled on a few years after that, and they did find an ore body. I think they actually made a little money a couple of times. They found what was called the Twinem Stope. It was, you might know the Twinem family over there, he was a geologist, maybe geologist slash mine engineer. He kind of plotted out this thing for Brad Place, that was the guys name, he was the main stockholder, remember him? He pretty much told him where he could find some gold in there... hit a good ore body, and he did. So I worked there at the start of that development, we did a raise and that was a long raise. It went up like 150 feet from the level we were workin' on. And found that ore body back there, and then they did what they called a slusher-drift system. I worked up there on that. It was real spotty ore, but quite a bit of it.

You know how slushers work? It's not on track, it's all done with buckets, and tuggers, and draglines. Down into an ore chute, where it would go all the way down and a track system would be and it would be trammed to the station and dumped. You didn't develop any kind of track system up there, you just did it all with slushers, draglines. Just get it out and down the hole, then it'd be picked up on a railroad system down below ya. A hundred twenty-five, a hundred-fifty feet, or whatever it was, it went down an ore pass. Where Genie Leaf would be down there. He got days pay which would be roughly what we got per hour, plus twenty-five cents a car. He'd take his twenty-five cents a car and put in a jar at home. You know that old truck he drives around now, it's got a camper on it? He bought that truck with the quarters he put into that. Course, trucks were cheaper back then. He saved the money, set it aside, bought that truck. He was a hand-tramming maniac, 'ol Gene. That boy's a maniac period. Hell of a worker. He was my partner, too, at the Ajax. Gene and I did a couple slusher drifts there, and did a long hole program for one of the companies that had the Ajax.

What is long-hole?

You know where you thread the steel way out there and take samples? As it comes back, take samples of it, and then it's all cataloged by the footage. You just thread the steel together, they come in four-foot sections; you thread it together and it's like a drilling machine. You'd just keep sending it out, and you're pushing water through it, through the drill, and you have the slurry that comes back out of the hole, and you have these little bags, and you catch the slurry and catalog it, and each hole would have a number and the footage thing, and you'd just fill that in, 640 x 145 deep, you know.

What kind of hoist did the Mary Nevin use?

That was electric. Commander Drum was one of the hoistman. He put Greg and those guys in the sump once and he just said, "I don't know what happened, I just forgot where I was." It could have killed them, of course. He went past the bottom station and right down into the sump at full speed. You know, it's got all kinds of safety things, but most of it doesn't work too well. A lot of times it doesn't. Then "Boom!" they hit the bottom and it knocked them around pretty good. I don't think anybody got seriously hurt. I wasn't in it.

I've heard you could tell the different hoistman apart by the way they ran the hoist?

Oh yeah, little fine touches, for sure. More careful. Yeah, just the way they come into the station. Some just come gliding right in. You know, you've got a mark up there where they come gliding in, all nice and smooth and boom, they're stopping just as you're, because they're waiting for the signal, too. One bell. Bing, and same time, they're stoppin'. They know where it is up there. There's definitely better hoistman than others.

Hoistman has a certain code. You had to have trust that they were gonna go by that code.
Well, they'd never move the cage without a signal. Don't care, no matter what. Never move the cage without a signal. Somebody might say, some hoistmen might say, "I've waited too long, I'm gonna move it." But a good hoistman never would move that cage. He'd wait there all day. Say you got off the skip; you're supposed to give a 3-3. It's a one, two, three; one, two, three. That's the release bell. Okay, we're off here, you can take it. Every once in a while, you're talking to your buddy and you get off the skip and you wander off, and the skip is sitting right there at the station, like this, and you get out, and you wander off, and you forget all about that release bell. Well, the hoistman, you know, he can't move that cage. There would be a hoistman who'd say, well, shit, they know exactly what's happened, they've walked off and forgotten about it. Some would sit there all day before he'd move that hoist.

What type of blasting material did you use?

I used all different kinds. When I worked at the Gold Bond and the Mary Nevin, we had that simple old orange fuse. You know, it burns slowly and it's timed, and it burns nice and slow. Fifty seconds to a foot and you cut 'em all the same length. Make sure they were all going.

Dynamite goes in the hole. Then we'd use that stuff, ammonium nitrate, fill in the hole, which is another explosive. First stick goes in there, the orange wick comes all the way back out. What we call safety fuse comes all the way back out, each one's got a cap on the end, and you wire them all together with what's called a spitter cord. You light the spitter cord, it burns through each cap, and ignites that particular fuse, and they're all staggered on there... so this one would go a little before this one, and so on and so on. That's your delay. See, you light it and watch it burn through all twenty caps, make sure... and the fuse is wax covered, so it ignites it in the cap and you can watch the fuse, the wax start melting. That means it's burning. So you'd sit there and watch, make sure they were all going, then "Fire in the Hole!"

If for some reason one didn't go, then you had crimping pliers, you'd cut the blasting cap off, and you'd take a lighter or a match or something, start it by hand. It burns at a real steady rate; you can see it. If one ever took off running, I never saw it. The main thing with handling dynamite was the miss that was left in the face. When it didn't go off for some reason. Say you had them all lined up, and the holes were like this, well this one goes off before this one and in that split second, cuts that fuse of the one that's supposed to go off right after it, so it doesn't go off. And it's left in there, and of course after a blast it's all covered with mud and dust, mainly. So you may not be able to see it.

The first thing you should do after a blast is you should get the water hose out, and hose everything down. Hose that face down, because most of these mining injuries around here, when you talk to the old timers, that's what happened to them. They drilled into them. They set up their machine, they didn't see a miss and they cross cut a live stick of dynamite and the power of the drill set it off, that blasting cap in there. Hits the blasting cap, sets that stick off. That's what happened to Doc; he was partially blind in one eye. I think that was from that.

Did you find missed rounds?

Oh yeah. With those orange safety fuses it was fairly common, enough that you would check. Definitely something that you would be real careful and look for. One of you would be barring down, and one would wash it down. The other reason you washed it down was you saturated the muck pile. The Ajax it wasn't so important, because there's so much water in the Ajax. But, water down the muck pile because you don't want to be breathing that dust, so you gotta keep saturating. So one of ya, the other's got the steel bar and is knocking down all the loose stuff. The integrity of the rock was pretty good.

What about blasting at the Ajax?

Well, we couldn't use very much prell, because it was so wet. Prell is real dry stuff and the high humidity and everything was so wet, you couldn't use it. It would just absorb the moisture. We used what were called non-electrics, and we also used electrics. But we just jammed the hole with sticks of dynamite. Instead of just running one stick into the back end of the hole and filling it up with prell, we'd take these sticks and one of us would just feed them into the hole. You'd break the paper by going like this so it would smash easier, and he'd set 'em in the hole. The other one would run 'em in, pack that hole full of dynamite. Because of the water, prell wouldn't work. There was just a lot of water in the Ajax. Especially below the Carlton Tunnel. The Carlton Tunnel drains the Gold Camp, you know. The Ajax is the deepest mine; it comes in at the 3100 level of the Ajax. It's also the bottom of the Portland. But then, the Ajax goes down another 250 below that, and that's where the water is.

Is that where the Bobtail vein is?

We were looking for it down there. We actually found it and followed it for awhile, but it had been mined on the upper levels, too.

I've heard if you surfaced at the end of the workings, you're at the Victor Cemetery?

Let's see, you'd get off the skip and head to the right. You used to get off on the 3300 and you'd be facing south then you move to your right. Then you'd go back, but not very far, and it'd curve like this. I'd say west, then you go south, and it seems to me you go east then, back this way.

Is it hard to keep your directional orientation underground?

I never did very well. I could always find where I was going down there, but I never thought in terms of east and west. No. Directions are so related to where the sun is, and that, on the surface, but underground... It was the engineering crew, the surveyors who knew.

Were there accidents where you worked?

Climax seemed to have the most, just because of the volume of it. So many people. So many uninitiated people. There was a young woman that was killed over there while I was there. I didn't know her, but rock fell...

Over here, you know, Ted almost lost his life one time. He was the foreman over at the Mary Nevin. We were driving a cross cut on one level, what level, I'd say that it was 300'. Anyway, we were driving this thing and it was called the Old Gold Cross-Cut.

They were trying to find the Old Gold, which was another mine; they were trying to find this ore body back there. Instead of taking the waste, dump rock, to the shaft and hoisting it up, we took it to this old stope and we dumped it down there, and had built a chute down on the 400 level, so it would all go down. It was a stope that had an angle like this, had a real smooth footwall. The footwall of stopes are like this, and have an angle. The footwall is the one on the bottom, and the hanging wall is the one on the top. The footwall had a pitch to it, and down here was a chute. We dumped up here, then it went down here and through the chute.

Now why did we do it that way, when we could'a just trammed it to the shaft. There must'a... I can't remember why. I think it was, well, I think it was a place where we could store it and we didn't have to be tying up the skip all the time. Anyway, the footwall had an angle; it didn't have a very good angle. What would happen is you would dump up here and it just would kind of hang up there, so it'd hang up here, then you'd have another big space here to the chute.

Well Ted climbed down from here and he was going to get this stuff loose. Was going to bar it loose. Well he got it loose, much more than he wanted, and it came down on him. It all came down and he was buried. There was somebody down below, you know, realized what had happened. Took them about, I don't know, five or six hours to get him out. Nobody knew if he was alive or not, because he was underneath this muck pile up in this stope. So the only way to get him out, it wasn't on my shift, was to pull all the rock outta the chute and hopefully here comes Ted. Finally, here comes Ted. It was big enough rocks that it didn't like smother him. Big enough there was air passage through it. "Say hey, here comes his boot!" Then they pulled him outta the chute. I don't think he had any broken bones; he was scared as hell. He thought he was gonna die for sure, because even though he could breath, he was under the muck pile.

Gary got hurt pretty bad up at the Ajax. He had the worst nightmare, because of course if you're gonna get hurt in a mine, it's gonna be in the shaft. Because it's the one place where you have no control. Once you give the signal to the hoistman to haul you up, there's no way you can stop that thing. In the Ajax it just flew. When that thing flew, the timbers just went by. Well, when you're down here, you get on the thing. What if something happens? This is a 3300' shaft, what if a timber just moved over a little bit or something, you know, and that comes up and hits it, you know?

So, anyway, if something's going to happen to you, it's going to happen in the shaft. I think most wooden shafts, like these old time shafts... you just don't have any control over it, what may happen. Because you can't stop it. Anyway, what happened is you know how the skip rides in the shaft on the guides, and how the guides were here and the channel liner fits over here, it rides like this.

Well, on this channel liner that's welded to the cage, there's shoes in here made outta steel. And that's what actually, you're on the side of this channel iron, smooth pieces of steel, and that's what it rides on. So, what happened was one of these shoes had busted off and got wedged between the guide and the channel liner. So Gary, I don't know if it happened when it came down, maybe it busted loose. Gary gets on the skip, I think it was on the 3300 level, and he gets on, gives the signal to go to the surface, and the hoistman starts pullin' on it, and it's wedged in there.

So what happens is he can't feel this up there, because it's 3/5s of a mile of ropes. The skip doesn't move but it stretches the cable, it stretched to a certain point and it busts loose. Just from the torque on the cable it busts loose, and the thing just shoots up like this. Shoots up, maybe fifty feet, just "woo"... shoots up and now this thing is loose. And you know, it's got a bonnet on it, and Gary goes flying into the bonnet. Boom! Hits the bonnet then falls back down in to the skip. It's a good thing he didn't fall, fly out and go down, you know, into the other compartment of the shaft.

So it's like a yo-yo thing, stretch – boom, stretch – boom... does this three or four times. Gary, he wedges himself, it all happens within seconds of course, Gary wedges himself, but he's got a broken ankle so it's real hard. Breaks his ankle on the first one up. He wedges himself into the bottom of the skip to stop himself, because for all he knows this is gonna go on all the way up the shaft. Well, the hoistman realizes there's something goofy. I don't know how he... something on the feel of it, goofy. So he stops it, and Gary was able to get over there and give it one bell. Stop! Don't move this thing.

The hoistman didn't know?

You know, I always think some would'a known, could'a felt that torque in that first one. I would think that this guy would have known, but nobody seemed to blame him. But he easily could'a killed him. The shaft is inspected all the time, too, but it was a huge shaft.

What about pranks underground?

I can't think of anything that would be really exceptional. Just the general grab ass. Oh, on the inside of the helmet of the hardhat, take that nitroglycerin and you put it on the band, on the inside. I heard of people doing that, but you know, it would have been too cruel of a joke because that's nitroglycerin based explosives. I was subject myself to getting headaches from that crap, anyway, and that'd be just terrible. I heard about that more in lore than ever saw anybody do it. Because anybody that had a powder headache knew that was, you know, it could make somebody deathly ill. And that stuff would make you so nauseous, such a powder headache. For me, I'd get such a bad headache that I'd end up getting nauseous, too, like oh, I just felt terrible.

A few times I couldn't keep working. The one thing about working in the mines, if you didn't feel well and you just didn't feel that you could do your work, your partner would always say, "Well, take five. I'll cover" It was always like that. Like most of the jobs that I had there was a certain pride in getting your work done, and you didn't want to go home if you were sick or didn't feel well or you had pulled your back the day before from lifting. Your partner would say "You take it easy, we'll get it done." That was always nice.

Did you ever work contract?

Yeah, I worked a little bit of contract. We had a contract system over at the Mary Nevin. We got a guaranteed day's pay, then we were paid by the foot also, over a certain amount. Then Genie and I got guaranteed day's pay on the long-hole, we got paid, I forget what it was, but if we put more footage in a week, over a certain amount, we could make pretty good money. We didn't take breaks. We'd get off the skip and we'd be gone. One of us would take five and eat lunch while the other one was working. The two of us never stopped working at the same time. Your lunchbox was right there.

Did you blast once a day?

Generally in the mines, other than taking care of slabs and misses, and maybe a slab round or something, the time to shoot was at the end of the shift.

We had a thing at the Gold Bond with Al Goodwin. You get your round in and I'll pay you for eight hours. This was like the only gypo he'd give, you get your round in and out, and I'll pay you for eight, and if it takes you six, that's fine, I'll pay you eight. So that, in a way, was a little bonus. Well, I worked with Ron Myers, and he was just a hell of a miner. He was my age, but he had done it since he was 15 years old, and came from a mining family. We'd get it in five hours, he could just really go. Well Al was a little bit screwy with the money, and after a couple weeks of that he'd say well, I want you guys to go, I can't be doing this, go back down there and get another round in. So we immediately quit that, we spread it out to eight hours.

The practical jokes, there were some. Ever see Ripley's The Trammer's Guide to Working in the Mines? It was done in the 1920s and it's full of those kind of things. "What to do with the new hire?" He was a real good cartoonist. You know Ripley's Believe It or Not? His brother worked in this camp as a young man; he did a lot of neat drawings of this area.

Were there beaver in the Ajax?

That's true. I never saw them, but I know Gerry, he worked there. He came runnin' back one time and said "You'll never believe what I saw back there!" It was around the corner from the Carlton Tunnel, and this is six miles from the portal. Said he saw a beaver. In the mine. Yeah. It climbed outta the piss ditch, you know, it's a pretty good stream. He swam all the way up and there he was in the Ajax. Scared the hell outta 'ol Gerry, because he wasn't sure what it was. You know, you got this thing swimming in the piss ditch and it just scared the hell outta him. Beavers are pretty good size.

Did you go down the Carlton Tunnel?

Um hm, we used to work down there. Because there was a sand bar about half-way down, and just something about the way the water swirled there with the pond of sand, it had to be worked on all the time to keep it open. We'd get down there on a diesel motor, called a Plymouth diesel motor, a locomotive that we'd take down there. Matter of fact, Gary Horton and another guy, their duty was with the best job in the mine. That was just to work in the Carlton Tunnel. So they'd come in, punch in at the Ajax, get in the company truck, drive down to the Carlton, then they'd spend the whole day down there. Goofing off, mostly. They had a backhoe that sat on rails there, that you could dig the, keep the piss ditch flowing. That was their job, to keep the tunnel open.

There use to be, beaver would come down there and dam that thing up, down at the portal. Then the water level would rise, you could see it back up, all the way to the Station there on the 3100 level. So, they'd go down there, I never was in on that duty, but I think they even dynamited them a couple times. I think they would call, I would hope they would but don't remember, but I think they would call the Fish & Game, and they live trapped the beaver. I don't think they'd go down and shoot the beaver and blow them up, too. By this time they're probably a little more environmentally conscious. They wouldn't be killing some beaver. But those stories are true, about the beaver.

Did you ever have to help clear ice at the Mary Nevin?

Yeah. That was terrible duty. There was water running down there and it would get real cold and ice over, then the skip wouldn't go through. So you'd have to take axes, and it'd always just be colder than hell. Just cold. 'Course, it happened in the wintertime. And you'd take an ax and break the ice.

You'd work outta the skip, just start pounding away. Because it couldn't get past that. You didn't want to get the thing wedged in there, you'd just start going where the ice started. Think we crawled outta the skip and got over in the manway and maybe got below the skip and started working ice off the guides, where it was built up. I think everybody, as I remember, had to take turns doing that because it was pretty bad duty.

After the Ajax closed, I got a job over where they were doing a shaft at the bottom of Rampart Reservoir. They sunk a shaft a thousand feet deep. Then at the same time they were runnin' a, it was going to be a water diversion thing, and they ran a tunnel to meet there, and they were going to divert water outta the reservoir down to the Air Academy. I worked on that shaft. That was done with a sinking bucket.

How did you do it?

Just put a big sinking bucket down. We could get four men in it, and sinking hammers, you know drills that drill down. We'd get in that thing and then they'd lower us down there and they'd pull the sinkin' bucket up above us, it filled in a good part of the hole, then they put something over the top, too, so that nothing would get kicked down on top of us. I mean you're at the bottom of this hole, it's bad duty down there, because you've got no place to hide. We'd drill the bottom out and get back in the sinking bucket and go out. And they'd shoot it, and you'd go back down and muck it out. They'd take it out and dump the bucket and come back and get you, when it was all over. That was a nasty place, because the water would just drip down there all the time. Always dripping on your head, and down the back of your neck. Oh that was hard work. Sinking hammer was a lot harder than a jack-leg. Jack-leg you've got that leg that will balance and hold, and do a lot of the drilling for you. A sinking hammer is a nasty machine; like a jack-hammer with rotating steel on it. I was only there about six weeks. Didn't last long. They were already way down there when I got there. I heard about the job, Eddie and I went over there. They didn't hire Eddie, because they thought he was too old. But they hired me.

Can one walk underground from Victor to Cripple Creek?

That's the story I've always heard. I'm sure you could at one time. I'm sure that with all the workings that the mines had connected with the drainage tunnels... I don't think you could do it in a straight shot, either. You'd be down the ladders and up over here to this ladder and down again. I bet there's some old timers that probably could, would know the way. Maybe they did, I don't know. That would have been before all this mining they're doing now. Well, the Roosevelt Tunnel would probably be part you'd have to go on, you know, that's pretty deep. It's on the 2000 foot level of the Ajax.

Tell me about high-grading?

Well, there was only a couple times that I was in what you'd call high-grade ore. And one was at the Bobtail on the Ajax, and one of them was in that Twinem Stope, over here. The one that Brock and Jim, Jr. drove over at the Ajax, was called the Monument Dillon, and those guys hit some really nice high-grade. So, I would say that at the Ajax, there was quite a bit of it. I think that there were guys that had their little operations going, that were over and above and around the company's operations. They had little pockets, you know. If you know where you're looking, you know how the ore is around here. It's real spotty. You'd have a little seam like this, then a little place that's got all kinds of little Sylvanite and Calaverite crystals in it, and you know about it, but somebody else doesn't know about it. I would say that yes, especially in the mines that had a lot of gold in them, like the Ajax.

I know of some guys that had places and they'd just disappear, a couple hours at a time. They weren't napping, I know they weren't napping. I knew a guy that had a, he'd go and pan up at the... you know, from the 3350 we pumped everything up to the Carlton Tunnel. Obviously had to pump it, because there's a lot of water there, that's the only way you're going to get it outta the mine, is to pump it. He had a little thing goin' up there where he'd come work... where it come out... pump it out and it came out of a pipe up there right into the piss ditch of the Carlton Tunnel, which is right off the Ajax Station. He'd spend a lot of time there with a gold pan, where it come out. I'm sure he found something. I know another guy, I know he had something goin' back there at the New Market, somewhere. Off on the 3100 hundred level of the Ajax. Had a little seam somewhere, he would always wander off and you wouldn't see him. And he always went by himself.

If somebody wandered off by themselves like that, you didn't think anything of it?

Oh, I think we all knew what was going on. It was just 'Whatever.' (laughter) You know, they're crazy stories, but I'll tell you, there was a lot of high-grading at the Ajax. I saw, you know, the Bad Boys of Cripple Creek turquoise guys, Bad Boy Mining, they had a booth set up here on Gold Rush Days. They were selling all this turquoise and stuff, mostly turquoise they mined, but then they had stuff that said "Ajax Mine Bobtail Vein." I looked at that stuff; I knew exactly where it came from, I knew how they had gotten it. They had gotten it from someone who had traded with miners from the Ajax. I recognized some of their rocks, even. I said, oh, I know exactly the spot this came from in the Ajax. They were asking a small fortune for this stuff that had been made into pendants and jewelry. And the stuff that Brock and Jim, Jr. hit down there at the Monument Dillon, it was mind boggling. It was just beautiful rock.

You know, the miners would hear about somebody that had a really nice muck pile and they'd all be headin' over there. I remember I had a partner; J.W. was his nickname. He was my partner for quite awhile on that Bobtail Vein in the 3300 level of the Ajax, and we'd get some stuff, and you knew it was a nice muck pile. And they got wind of it, somebody did. A couple of these high-graders, I just won't mention their names and they're not around here anymore, they were on that muck pile. Finally we had to say, "Hey, you guys gotta go, we got work to do." And somebody, we built this crib raise. You had to have a secondary way out and a way for air to move through the level. So, the 3300 hundred level was a newly developed level. Well, once you got back a certain ways, according to mine safety rules regulated by MSHA, you had to drive something up to get some air flowing through there. So, we got back in there a ways and then we had to drive this crib raise up to the 3200 hundred level. Well, it was driven right on the Bobtail Vein, and you know, there was some really nice ore in that crib raise.

I worked with these guys and this friend of mine, Ron, he worked the shift before me, and he'd say, "Check over there on the left-hand rib when you get up, there's something good." This is all stuff, a lot of times, would be just between the miners. We just wouldn't tell the bosses. And you know how the vein structures work, you've got a main vein system, but there can be any number of little stringers and pockets off of it. A lot of stuff just got passed by, but the miners would know about it. So, I think a lot of the stories when the miners say, "Well, if I could get back down there, there's something only I know about." ... they're true. They may be exaggerated, but they're true. I can go out in the yard here, and I can pick up a rock and I could tell you, if it was a mine we could get to, I could take you... and this is 20-30 years ago, I could still tell you which mine and exactly where it came from.

Did you pump in air, even the Gold Bond?

Um hm, just run a vent bag back. Al's was a real gassy mine. Yeah, the Gold Bond. Since they started doing this open pit, a lot of the mines, this is my understanding of it, that opened up some things and made air flow change. But the Gold Bond was real gassy, and the Ajax was. There's lots of situations where we'd work in gas. You'd just put the lamp up a little higher.

Where the gas would stay there down here, it's fine down here. There's always some gas in real deep mines, there was just gonna be some gas, and by God you gotta get your work done. So, you'd work in certain gassy situations; just hang your lamp higher.

Hell, we'd do that on mine tours with tourists. We had a place in the tour where we had a safety lamp burning. You'd explain the safety lamp and we always hung it low, for our own information. But when it was gassy, we'd just hang it higher. And tell the people, "Pick up the little ones here, it's kinda muddy." (laughter) Yeah, anybody that's guided tours in the old days will tell you the same story. It would be a joke among the guys, telling them to pick their kids up. You literally would do that. The kids down there kinda breathing hard, and Daddy doesn't have the foggiest what's going on. But that's not a problem anymore. Everybody seems to think they've opened something up with that open pit, and the air flows differently in all the mines.

What about Tommyknockers?

You know, I think that would'a been legend. We know where the legends came from, and I just never thought about it much. Just jokes. I think that was something that we never really talked about. It'd be a joke or something, "Oh yeah, the Tommyknockers took your lunch, or took your candy bar out of your lunch." Something like that.

It strikes me that back in those days of like the 1970s and early '80s, until the Ajax closed, they were the very last days of that style of mining. It's still being done here and there on a small, what we call 'poor-boy' operation, but that track mining like that, small mining, that had been, when we were doin' it in these old mines around here, like the Gold Bond and the Mary Nevin, the Ajax... we weren't hand-steeling, but the mining hadn't really changed since the 1890s a whole lot. And that was kind of the end of an era. We were the last ones to do that style of hardrock mining. I don't think that type of mining will ever come back. Track mining, small scale track mining. I think everybody that was involved in it was pretty proud of bein' the last ones to do it in this Gold Camp.

What was it like the last day at the Ajax?

I was workin' in the 3300 level with Jim, Jr. We were in the Bobtail Stope. We had already stoped one side of the cribbed raise, it was about a hundred foot long stope and a hundred feet high, and we'd done that over those previous couple months. And then the Bobtail was driven further back, another hundred feet and we just had done the lifts, the raises for the chutes for the second phase of the stope.

What you do is you run a raise up like this and maybe another twenty-five or thirty feet another raise like this and they're all timbered. We had three or four of them like that, all timbered. So you have these raises that go up fifteen feet, then you join them all together by driving along the vein right here; they're all joined together. See what I mean? Now you're set to start stoping, and you start going up, drilling in the vein and you start excavating it all down through the chutes. Well, we had gotten all this work done. The chutes in, the raises in, that first cross-cut that joined 'em all together.

We were up there with our stoper. He was down in one end of the stope and I was on one. We each drilled about a hundred six foot holes with the stopers, right along the vein. This was going to be the first shot. And, I don't know, went back to the Station to do something, back for lunch break or something at the Station. Anyway, the phone rang and picked the phone up and somebody said, "Hey," it was like two o'clock in the afternoon, "Hey, Ed Hunter says get all the hands up on the surface." We knew there was something wrong. So, we left.

We got on the skip and went up and he said, "Well, Texasgulf is shutting the place down. Hecla's taking over." At that time there were about forty people working at the Ajax, he said, "A bunch of you guys are going to be laid off. Hecla will keep some of you." And you know, that's it.

Well, I was one of the ones they kept, but it was about a two week transition period. Then we went back to work. But we never went down, when we went back to work; they were there one more year. We went back, but they never went down to the 33rd again. Those holes that me and Junior, those 200 holes in that ore body, they're still sittin' there, under water now. They never went again. They drilled that Monument Dillon with Brock and Junior on their real high grade. I worked on the long-hole program. Things just... I don't know, they just didn't look good. Because we all knew if you didn't go down there on the 33, you know, that's where the gold was.

So one day the President of Hecla came and he said, "Things don't look good. If we don't find some ore body here in the next two weeks, we're gonna have to shut down." Well, there was a mad scramble down at the end of the one that Brock and Junior were doing. They went off in all directions looking for some high grade. Had me and Genie down in the Carlton Tunnel running the slusher drift. It was called the Jenny Mae, was the slusher drift. Gene Leaf and I knew, it was like down to the end of the two weeks and they'd gone off in these different directions. Set the last round off. It was the last round set off at the Ajax, was me and Gene. And we knew it was over. That was it. So, they had two kinds of hints. When they quit that Bobtail Vein down on the 33rd, we knew that, well, they're not gonna be too serious, or something's wrong here. I don't know what the deal was. They were here a year. I think there kept a couple of guys to help get some of the machinery outta the mine. Although, they left most of it down there.

They thought they'd open again?

Well, they thought they were. They just didn't care; it wasn't worth the effort to get it out. Then we were all up there on the surface, that last day. That's the one I said I wished I'd of taken a picture, because we all knew. Romey was this nice guy, he just wanted to pep us all up, he says, "You know, it's only gonna be a couple 'a weeks, I got the inside word, we'll all be back to work." I never saw him again.

I used to work swing-shift, and I could tell exactly what was goin' on at the mine by how much rock was comin' down that skip. You know, they'd bring it up and dump it on the surface. You could hear it in the big steel chute. I could say, "Man, they didn't get it all mucked out." I could tell, always, what I'd have waiting for me at work by how much activity was going on in the morning at the mine. You know, when the mine closed and that, you didn't have that sound any more. It was kinda like loosing part of Victor. It was tough on Victor. When the Ajax closed, it was tough on Victor. The two things that happened in those years that happened that was tough on Victor was closing the school and when the Ajax closed. Like the school, it was so neat in Victor to hear the kids on the playground. The life that brought to this town, and then Bingo, one day it was all gone. Both. Well, Sylece, you went to kindergarten there, didn't you? You and Anna were the last kindergarten class in this school. But it was so neat, you'd be up town and the kids would be on the playground, and it'd be so neat. You could hear their voices in town, no matter where you were in town, you could hear the kids having fun. A lot of people left when the mine closed.

When did the Ajax close the last time?

Somewhere in '84, '85. It was tough on the town. You know, places like Zeke's Place and that, the mine brought action in. And they just left. A lot of them just had come here to work at the Ajax. From Leadville and other places, "We should work at the Ajax Mine." That was the same time Climax was closing. So a lot of Leadville hands were here.

Then it was like, I wasn't going to go any where, and neither were a bunch of the other guys, and so we went from having this job that we took a lot of pride in, and at the Ajax we made good money, had good benefits, you know, it was pretty decent money. All the sudden you didn't have a job. Then they started going through all the old dumps. TG started with this experimental pad over there, the very first pad. So they called up us hardrock miners, that worked in the hardrock mines, now we were Pad Monkeys. We went from making really good money to just barely making any money, and being nothing but shit birds. Yeah, so that was like, oh man, that was a bummer. It was just a bummer for all of us, although we all had pretty good senses of humor about it.

We realized, you know, we went from this job where we took a lot of pride in it, and it was hard work, to having this job that anybody could do. It was ridiculous and the pay was low and you went from working at the Ajax and getting full benefits and vacation and good money, to no benefits. You know, it was kind of like a kick in the ass. And they exploited it, too. They knew that there was no work, so they paid as low as they could. But I ended up working on the surface, and in surface mining for quite a while. Another ten years. I ended up being in the metallurgical end of it, in assay. You know, the pay did get better.

It seems people that like to go underground always like to go underground?

Yeah. Then there wasn't just any other, I don't know who works over at the surface mine now, but there's only a handful of us that had worked underground.

We had the C-7 project at the Cresson, with Ed Hunter. That was a hell-hole. A big can. Was right next to the hoist house at the Cresson Mine. It went in to the side of the hill. It was kind of Ed's project; he was in charge of it. We'd go to the Ajax and get in the truck and drive over to the C- 7 and work, then go back to the Ajax. They just had a low-grade body of ore there and were taking it out. But it wasn't a tracked mine, they had a scoop-tram, and Young buggies. The scoop-tram and Young buggies that we used on that are parked over here across from the mill on Highway 67. From that C-7.

It was disappointing, I don't know why. You could never really see the ore. I think it was so close to the surface that it was heavily oxidized and it changed the rock a different color. I don't know if there was any... I think it was disappointing to the Company. It was a hell-hole to work in. Colder than hell in there. It was winter when I worked there. But it wasn't track, it was all slusher drifts. Had a scooper tram to get in to draw holes and it was picked up by the scoop tram and thrown in those Young buggies, and hauled off and dumped on the surface. Some of it may have even gone to the mill; I'm sure some did. I'm sure there was some gold in it.

I think there was some problem with other claims involved, and there'd be Texasgulf had cut the pie up too many ways. That was a problem at the Ajax, too. That old Carlton, I don't know if you remember him, he's passed away now, but had to be close to 80 years old, and he owned a certain share of the Bobtail. One of the Carlton's, he was like 80 years old. He'd come down there and climb up in the stope on the 33rd level of the Ajax, he'd come down. Just checking it out. But he obviously had some kind of percentage in the Bobtail, either that, or maybe he was involved with Golden Cycle. He was a real nice guy.

Was the Ajax deep water on it's lowest level?

Yes. You'd drill into water courses and it'd throw water out. Couldn't even load the hole. If you had a hole here with water pouring out, you drill another hole next to it. You just had to drill more holes. Sometimes you still couldn't load them, there'd be such pressure coming out. More holes 'til you could get something to stay in there. Hopefully it'd stay in there.

Ever find carbonated water?

I don't remember carbonated water; I'm sure it was there at the Ajax. I do remember gas coming in with the water. Yeah, coming right outta the face and you'd put your lamp right there by it and it would literally hiss. It's like you released something that was held back there in pressure, and you released it with the drill hole or something and 'Ssssss', put your lamp there and it'd put your lamp out. And the water would be coming outta there, also.

The Moffat was another hell-hole over there. You know Chuck Dager; you remember him. He hired Ron and me, and we got a guy named Kevin to help us portal the Moffat in. We portaled that Moffat in. There had been a big cave-in back about a thousand feet into the tunnel. He hired us to clean the cave-in all up and timber that. So we did that job, took us most of about nine months to do it all. Then we go into it with him about pay, he was another guy who promised you something but it wasn't in writing, then when it came time to do it, he wouldn't do it. So, we ended up quitting and somebody else carried on after us.

We portaled it in. It had all caved in. So we built a big portal, timbered it all up. We used these huge timbers; I don't know how in the heck we ever moved them around. They were actually timbers that had been part of a railroad bridge that old Chuck got. They were just huge timbers. I'm sure it's still holdin' the rock back, because it was a massive job. I don't remember how... it was with bars and whatever, I think we hooked up some kind of boom with a come-along; moved the stuff around. They were like 17 feet long.

Ron and I were the first ones there. Chuck ended up getting in a tustle with about everyone. He was that kind of person. One time we had the footage and he said, "Okay, I'll pay you tomorrow." The next day he said his uptown buildings had been broken into, he accused us of rippin' him off and said that nobody would get paid until he got back what was taken.

I worked with Johnny Nothaus at the Ajax for a while; he was a character. He could draw really well. You know the only reason I know that is because down at the bottom, where the Portland shaft comes to the Carlton Tunnel, the bottom of the Portland shaft, I did some work back there, some long-holing with Gene Leaf. Get off the Ajax at the 3100 hundred level, walk out a ways, we were in the Carlton Tunnel, and walk back the distance between the Ajax and Portland headframes there, and there's a big workbench there. For mechanics or drill dockers or whatever, there was a work area there at the bottom. And on the workbench, on the wood, there were some drawings done and it was signed by Johnny Nothaus. Just these really neat drawings. I don't know kind of mountain scenes, done in pencil. And that's how I knew this, he could have been a cartoonist or something, really, he was gifted with a pencil.

What about your first time in the dry in your shiny clean clothes?

Oh yeah, you could tell the rookies by all their new gear, for sure. I don't remember any real impressions for myself. I do remember being impressed by how you could tell who was the new hand. Brand-new miner's belt. I remember the dry at the Ajax the most, because that was the classic dry, where you hoisted your basket up. I don't remember that much about it. I remember how dirty you would get. The Ajax, up here, it was a big step up because they had showers, and a dry. Where the Gold Bond and the Mary Nevin, you know, I'd come home from the Gold Bond wearing what I wore in the mine, and I'd be filthy dirty. And the Mary Nevin, we changed our clothes, but there was no shower. At the Ajax they had hot showers, big showers.

I remember, coming outta the Ajax, what people would do, they'd bring their winter jacket and put it in a trash bag, and you'd go down in the Ajax... because in the winter you'd go from being 90 degrees and then just a minute or so later you'd be up, it could be twenty below, you know, up on the surface. You'd be soakin' wet, and you'd make a run for the dry which was just, you know, fifty yards away, but you'd be iced over by the time you got to the dry. So, a lot of guys, I didn't, but they'd bring a trash bag with their winter coat and put it on when they got on the skip to ride out.

The temperature at the Ajax drastically changed when you passed the two thousand mark, because that was the Roosevelt Tunnel. When you went below the Roosevelt Tunnel the temperature warmed up significantly. It jumped up from maybe 60 degrees to 75 or 80. But then as you got closer to the surface, it'd start getting cold. But it went so fast, you could be down in the 3300 hundred level in just a couple of minutes. Like I said, the timbers went whoosh, whoosh, whoosh.

Independence Mine

Beans on the Table

"The man said, "You will," and I said, "Yes, Sir."
~ Edgar T. Hunter

I was born in Yonkers, New York, in 1926. My father, Edgar L. Hunter, was a stock trader, and my mother, Catherine, was a stay at home mom, and later a private secretary. Very soon after I was born my family moved to Philadelphia, and I was there until World War II, when I finally got to be 18, and enlisted in the Army Air Corps. I was in the service for a little over two years. By that time the war was over, and I went out to Kwajalein in the Marshall Islands, during the bomb tests at Bikini Atoll. I was a radar repairman, so all I did was sit on Kwajalein, and when the B-29's would come in we'd fix their radar if it needed fixing.

Then after the war I worked a little bit, because I didn't know when I could get out of the service. I worked in a laboratory. It was the only job I could find, frankly. Near my home; it was within walking distance of a mile or two. Then I went out to Golden, when I was able to start the semester out there, and went to school at the Colorado School of Mines.

Fortunately the GI Bill permitted me to stay there. I did leave school a year because I ran out; I had used an extra year because strengths of materials and I did not agree too well. I worked a year down in Uranium country for the head of the School of Mines mining department, Clifton (Curly) Livingston. He was trying to develop solution mining of uranium. He had a little claim in Slick Rock, Colorado.

This was during the development of an underground block caving copper mine, and some contract drifting. I worked for a fellow who had graduated from the School a couple of years before, Russ Wood. We sank a little shaft and put some holes into a pod of uranium ore. We tried different solutions; pumped them back out, and then they sent them up to Golden to be tested. It was too small of an operation, and at the end of the year I went back. Cherry and I were married, and I worked as a hod carrier for the summer at the Water Works in Golden. I worked one summer between semesters at the Climax Mine.

After I graduated in 1953, we went to Oracle, Arizona. I worked at San Manuel Mine for a couple of years. This was an underground block caving copper operation. I worked underground for a while doing highly technical things like putting up pipe lines, laying rail for people who were actually mining, and some contract drifting.

Then I went, not real happily, to the Surveying Department. The man said "You will" and I said, "Yes, Sir." This is a big operation so they had this all laid out on maps. And we had to lay out all the drifts as they drove them. These are big, 10x9', at minimum. We put those in and then they sank three more shafts while I was there. They had the one main shaft that we worked in, and the three new shafts, and my job with my boss was to keep them vertical. The miners would sink about 36 or 40 feet and put in steel sets. We had little plumb lines for them, which we set so that they would line the plumb line with little dimples that we had put in the steel sets, so that they would line up vertically. And that meant they had to block them, and square them.

At one shaft, the old-timer boss made them put each set in right on line, or as close as they could get it. I'd go to the shaft and check those sets, and all the work had to stop while I was checking them, which aggravated these people who were on contract. I'd be there an hour, an hour and a half, maybe. Max. Because they were all just about right, just took a little bit of adjustment. Driving some wedges in one corner of the shaft, or another.

The other two shafts, they were twin shafts, you might say. They were parallel to each other; if I remember right they were about five hundred feet apart. Those guys would just slap a set in and blast. Because that's where they made their money, by contract. Then would come the day of reckoning, and I would go to the shaft and we would line the sets up. It takes a lot to move a heavy steel set with driving wooden wedges.

I'd been down there with the crew as much as two full shifts, trying to get those six sets lined up. Rather than wooden wedges they used their ax heads, which were steel. They made real good wedges, but shucks, by the time they ended their shift they were down about four sets. They ran out of axes, and they had to wait while the one ax was passed from corner to corner of the shaft. Why they didn't make some steel wedges, I don't know. Or why the boss didn't make them line the sets as they went.

I wanted to get more into mining and less surveying, and we moved up to Lark, Utah. I worked underground and did some drifts and stoping. I had done some in San Manuel, first, before surveying. And had worked both company time and once in awhile a contract crew would want an extra hand, so that was a help. But in Lark it was all company time. We did the drifts and the stopes. Then I was sent over to sink a shaft. At San Manuel the shafts were being sunk with a big Jumbo that went down, and the miners could take their drills off the Jumbo with arms, and drill the round out very quickly. Then load it, and then after blasting, muck it out. They had what they called a Ridell mucker. This was a little cart that ran up in the last steel sets. It could journey from one side of the shaft to the other along rails in the last set. It had a clamshell underneath; the bucket was let down to the bottom, and the clamshell would load it. Well, that was pretty good. It went pretty fast.

When I went up to Lark, Utah, we sank an old shaft further and it was all done by hand. We had jack-hammers, and we drilled the round out and then we would blast, and then we would use #2 round-point shovels to load the bucket. It'd be hoisted out and they'd bring it back. It took us about three days to get the whole round mucked out. Then we put in a timber set, and start all over again. So, it was quite a change from Arizona, I assure you.

The boss would come down about once a day and he was about as deaf as you can get, because he had worked underground too long. He said, "Oh, machines never bother me," and I believe they didn't, because he couldn't hear anything. He would make a routine appearance and then leave. We had a couple of real good shaft hands at that job. It was single-shift, which was very handy.

They brought some miners in from Eureka, Utah, for a second shift. At that time the man told me I was to come to the Engineering Department at Lark, which I did and I spent too long doing Time Study work for the mine. They would time miners in driving raises, or drifts, or stopes, and we would then compile a bunch of average times. We would try to figure out a new cycle for the work that would be more efficient. It took a lot of time, work, and paper, and it really wasn't very efficient. Most miners had worked it out to be about as efficient as they could get under the conditions and the equipment they had.

One of the big movements at that mine had been just before I got there, they had used the old-fashioned, great big, heavy drifters. They would put up a steel column and they blocked it in and then the drill machine would be fastened to an arm on that, and they would drill the holes from that. They were very heavy, maybe 300, 350 pounds. They switched when the Swedish people brought in what they called the jack-leg drill. It was a jack-hammer that was on a pneumatic leg, so that the leg would move it forward. One man could handle that pretty well. Much, much faster; way more efficient.

The mine pointed out how much the rate of advance had increased after they had rearranged some of these schedules. Well, that was a bunch of baloney. It was all this new jack-leg drilling, as opposed to using the old column and bar, heavy drifter. But, they thought that was pretty good, so it worked out fine. That was a big mine. We were around, and under, the fringes of the Kennecott pit at Bingham Canyon, and the Bingham copper mine was in a big stock. We were mining in the beddings that sloped into that stock. These were limestone beddings that had lead and zinc in them.

Before I'd gone there they'd had a shaft up in Bingham. I don't know if you have ever been to Bingham Canyon, but now it's pretty much of a guided tour, keeping cars under control because they are so extensive in their mining operation. But at that time you went up, you actually had to go through a tunnel like the one that used to be on Highway 67, to get to the mine office. Then you'd go in a little adit, or a little tunnel, and down the shaft.

Kennecott wanted all these hands away from there, for that was interfering with their business, with their operations. So they built a three and a half mile tunnel for the old United States Smelting, Refining & Mining Company, for whom I was working in Lark. They built a new change room for them, and office, so that they could all have the change room in Lark and go in by man-train to one of the two separate sections of the mine. One was called the U.S. Section, it was nearly vertical. The beddings were pretty near vertical and the Lark Section was on an incline. The beddings were on an incline as they sloped into the Bingham stock. There you'd go in a mile on a man-train, walk to a shaft, get on the cage and go down either 500 feet and then walk back to an incline and take a little skip down the incline you were working on.... or you could go back to the original shaft and go down to the 1000 level, take the train back to another incline, and then get on a skip and go down to your level. The rock that was mined had to come back out that way, so it was pretty well worn out by the time it got to the surface.

At that time they would load hoisted muck at the shaft; into big mine cars. I say 'big', at that time they seemed big. They held about ten tons. The car would go to the yard at the portal, and they would dump it down to the big railroad cars underneath. These, then, would be taken about ten miles to Midvale where the mill was, and they'd be processed there. The company had had a smelter a few miles away, but they cut that out because it was too much problem with smelters being in the outskirts of Salt Lake City. This is pretty early and they kept records of egg production, and milk production and all these kinds of things to show that the smelter smoke really didn't effect the animals, and therefore it shouldn't effect the people. They were wise enough to discontinue the smelter; they shipped their stuff quite a few miles away to the other side of the Oquirrh Mountains, which kind of bound Salt Lake City on the southwest.

After a while at Lark, I was transferred to New Mexico where they opened a new copper operation. A place called Vanadium, just outside of Silver City. We had a contractor working at the Princess Mine, and one at the Continental Mine. We lived at the Bullfrog Camp; about 20 houses in Vanadium. Across the valley, at the foot of the Santa Rita pit of Kennecott was the Groundhog Mine of Asarco's. The Princess Mine was kind of in-between.

Then the company started a new operation up at Fierro, and I was transferred there as the Development Superintendent. It had been mined by the old Hanover Bessemer Company for iron many, many years ago, and had shipped that iron up to C F & I in Pueblo. But that got to be a pretty long haul, and I don't think the grade was enough to satisfy them. But UV Industries, formerly USSR & M Co., sunk two shafts. Boyles Brothers was the contractor. When they were finished and started developing the mine, then I went up to that operation instead of being associated with the contractors down further south. We developed the mine, and started stoping.

Did you do engineering?

Yeah, well, I was doing all engineer work with the contractors. By this time I was the Mine Superintendent, and then was the General Mine Superintendent, because the mine took over the old Hanover Mine of New Jersey Zinc Company, which was at Hanover, New Mexico. New Jersey Zinc pretty well mined out all the zinc in that operation. However, we did make serious attempts to work on it.

At Fierro we had the Number One and Number Two shafts. The Number Two was about a mile from the Number One shaft. The mill at that time was at the Number One shaft. We had an underground haulage way on the 1300 level that went over between Number Two and Number One. One contractor was at the Number Two shaft, and he was responsible for hiring all of his people, paying them and then the company paid him. Then at Number One shaft we had three different contractors on different levels. They hired and fired their own people. That worked quite well. We were mining mostly chalcopyrite copper ore, and it ran almost two percent because there was some bornite in it, which is a higher grade copper.

But there was a lot of pressure from the mine Union, particularly the people. They had all worked at New Jersey Zinc and one other company that had shut down. The contractors up at Fierro and their men, particularly, did not want to hire those fellas who thought that if they drilled about eight holes in an eight hour shift they were really doing quite a bit of work. The fellas at Fierro were all on contract and they couldn't see that slow pace. They were going to make money. So, they didn't want them to come up. A few of them came because these are people that lived with neighbors, and knew who could and who couldn't do a day's shift.

We also had quite a few people, miners from Mexico, who were really good miners. They were all on green cards; this is some of the hassle that's going on today. But the Union would get a hold of their Congressman, who would send the Immigration Service up. They would arrive at the shaft right as graveyard shift was coming off and dayshift was going down. They would check cards and decide that they weren't valid or something, and ship the hands off, back to Mexico. It was interesting, about three days later all the hands would be back. I don't know how they got back. But everyone was glad to see them, because they did such a good job.

One thing that I was interested in at the time, the State of New Mexico decided that all drillers should wear earmuffs, in addition, not just the little plugs. Well, you can imagine there was a lot of static about that. But, because the contractors and their bosses, the mine's bosses, like myself, and my boss, would go around, would insist that they be worn. All the sudden, after about a week, there was nobody who wasn't wearing their earmuffs. It turned out all the sudden quite a few people could hear better when they went home. They decided they weren't all that bad. But all the usual excuses, "I can't hear my machine, I don't know if it's working right." Blah blah blah. I think miners are resistant to change, quite a bit. This was true when hardhats came in, this is even with people who were putting varnish on felt hats to make them a little more resistant to something falling down. And particularly with safety-toes, they had a big campaign: "A safety toe doesn't weigh much more than a quarter." Well, put that on the end of your foot and see the difference. But it saved a lot of mashed toes.

Did you see these changes come through?

No, they were before my time even. I saw the earmuffs come in, and that was good. It's just like the boss that used to come down when we were sinking shafts up in Lark, Utah. You'd say, "Dick, we're going to need more powder next week. You better order." He'd say "What? "What? I'm not hard of hearing." But anyway, I did see that change take place.

Another one that I did not, thank goodness, see, was when the mines went from dry drilling to wet drilling. A lot of that was due to a man up in

Denver, George Leyner, who figured out how to make a machine have water go through it, if you hooked it up to a water supply. Then the drill steel had a little hole down the center so that the water would go through the machine down the drill steel hole and come out on the face, and dampen the dust, so that wouldn't get the clouds of dust that you get from dry drilling. Particularly if you're drilling with a stoper, where you're drilling overhead and driving a raise, everything came down on you. They were the ones you hear about, the Widow Makers, and fortunately as I say, I didn't have to go through that. But, I knew enough of the old-timers who had suffered through that, who had Emphysema or Miners Con and it's a tough way to go in their later years.

We finally ended up on a labor strike. There was enough push through the Government and other things that we had a strike, and they were forced to get rid of the contractors as contractors. So, during the strike they opened up the Alaska Gold Mining Company up in Nome, and I was transferred up there in 1975, as Manager. Cherry came up, and we lived up there in Nome.

We were getting the dredges back in operation after thirteen years. These are big. I don't know if you've seen bucket line dredges, but they are really a big floating washing plant, with a big snout, a big steel beam that sticks down into the water, and has a set of buckets on it that go around the long, steel, they called it a ladder. About a hundred-thirty-five to a hundred-thirty-six buckets on that. It was like a bicycle chain sticking down there, and instead of having the little round thing where each part of the chain was, there'd be a bucket. The bucket would hold about nine cubic feet of gravel and dig that and come up into the dredge, go through a big rotating screen, and hopefully little fine particles of placer gold would fall out into some sluices and they'd be collected, and then the rest of it would go on out the back, and you'd have to keep moving the dredge ahead as you found new rock to dig.

One of the problems was, you were in permafrost so the ground was permanently frozen. That had to be thawed before you could dig it. Fortunately back in about 1920 a fella had proved that you could use cold water, which would be like right from the rivers in the end of May, start of June. If you put enough holes down, drill holes, down through the area you wanted to dredge. Then you pump water down there, usually about a year ahead of time, not a full year, just during about four months of the year when the water was above thirty-two degrees in the river, and you'd pump that down under three or four PSI. The water would go down, it would depend on where bedrock was, but when we were there we were pumping water down about ninety feet. It would come back up around the pipe. As it came up around the pipe it would thaw the ground and that would slump in again, against the pipe, and then the water would follow the slumped material against the harder permafrost and thaw that. And in theory, in four months time, which is about all you had, you could thaw a cylinder of permafrost sixteen feet in diameter and the depth of the gravel. We had previously drilled holes on sixteen-foot centers. So that meant that each sixteen-foot circle, that was in theory going to be thawed out, overlapped the ones on each side in all four directions, so that you thought you were getting two cracks at it and it would be thawed out.

The big problem, oh, if you did that say this year, then you could come back next year and dredge it, as long as you thawed it well. You had to keep readings, and had thermometers that went down and checked the gravel and that kind of thing, to make certain is was indeed thawing. It would freeze back on the surface, but that didn't really make too much difference. You could mount a big hydraulic giant, you've seen pictures of the hydrayulicing of gravel banks where they have a pointed piece of pipe that is tapered, and it has a big head of water on it, so you have a lot of velocity of the water going out under pressure, and it just works away the bank. You can use that from the bow of the dredge, if you have too much frozen ground. Or, you could put short thaw pipes in and be thawing as the dredge advanced, because it's quite slow. It might go six feet in an eight hour shift, and then move the dredge over sideways if you had a wide ore body in there, and you'd mine maybe another 105 feet wide.

When I first went up there we were mining an area that was 330 feet wide. So we'd move the dredge three times to get that, and that would advance it six feet. The dredges could keep going, the management would say "Well, we want to go up to almost Christmas-time." Well, when you look through the historic records, one dredge one year had worked into December. About November 5th to 8th was when dredges, for many years, had to stop. Because what you'd do is take water from the rivers. There were two rivers up there, the Snake and the Nome. We'd use the Snake River water, because it was closer. Had a big pipeline that pumped it over there; it was 42 inches in diameter at the pumps. It went about a mile to the dredge and was much smaller when you got up there. We had little, they called them breathers, and they were pipes that came out from the pipe, with a little hook in them, like a cane, so that you wouldn't get stuff down the pipe. Then we had a little cup around them, and when it got cold we'd put some oil in there and some packing so it's be like an oil lamp, to try and keep it warm.

Well, we faithfully did that, but about the first week in November we were getting a lot of resistance on the pumps, so we thought we better check the pipe. We found out that the pipe, 42 inch pipe, had about a one inch hole in it. All the rest was ice. The old-timers up there told us if it freezes solid it's going to set up a vacuum with the pumps and you'll flatten the pipe, and then it won't be any good any more. So about two-thirty that morning we were out taking the pipe apart. Then that meant that we had about three or four more days to dredge, because you have to have new, clean river water coming in to effectively collect the gold. They said that if percent solids in the water got too high it does not collect the gold as it should, so you lose your product: gold.

So, it's very interesting, we were doing it the same way they had for years, since about the mid-1920s, except when we made the holes for putting the thaw pipes in we could use a modern drill, and drill the 90 feet in very little time and slap a pipe in. Where back in the late twenties and thirties and forties they had to either drive the pipes in, like you'd put a steel fence post in, and keep twisting it and beating on it. Or use a churn-drill, which was a circular bit and usually had some hard surfacing on it so you cut down into the ground. It took a long time and was pretty expensive. So the modern drill was about the only advantage we had.

Then we worked the next year, and we had a lot of people in the thaw field, as we called it, working on these pipes to keep them going because they would plug up. It was a real art to keep the water flowing. If you didn't have the water flowing, you weren't thawing the ground.

One of the big dangers in that operation with the dredge was that as you move the dredge forward, if there was a pillar of permafrost hadn't thawed out, that was enough to flip the bucket line off the big sprocket, just like a bicycle chain, and that could result in the whole dredge flopping over. And that was a pretty scary thought. So you wanted to do a darn good job at thawing. The second year we had a bunch of experts who had worked on dredges, and they said, "Well, these big buckets that you have..."

They weighed almost two tons a piece, and they had big steel pins that connected one bucket to the next. The bucket itself sat on the ladder and had eyes in the front and back of the bucket. These eyes were holes with heavy steel from the bucket around them, and then one bucket would fit into the next, and then you'd drive this pin, about six inches in diameter, and that would make the coupling. So that the buckets were all connected through the eye of both buckets. Everyone was of the consensus that we could go four months and the company had already negotiated with some foundries to make a new bucket line. 136 new buckets, it was quite an expense. But one of the problems was that OSHA had come in and shut down most of the foundries that were capable of doing that in the United States. So we had to get them out of Canada.

Well, nobody thought much about it because they figured, okay, during the summer they'll get them shipped up to the west coast, and they will come up to Nome by barge, and this is how we got all our freight in. With the big container ships, and you'd put things in a container, you've probably seen them, it'd be a box about 8x8 by maybe twelve or fifteen feet long. They'd just stack these on top of this barge, and then tow it up. They could tow it into Nome as soon as the ice got off the Bearing Sea, which would be late June. Then you'd try to get every bit of material you could into Nome. At that time, for the year. You'd have about three barges usually; three trips. You had to make sure, because otherwise you had to fly it in, which cost through the nose.

So, to make a long story short, three weeks into the season some of the eyes that hold the buckets together began to break. Then the big pins, the six inch diameter pins themselves had thick cracks in them, and we were afraid we'd lose the whole bucket line. That we'd lose the dredge, if it hit right and flipped over. So they started calling in all these buckets that they had, from all over Canada, got them out to Seattle, but there was no barge coming up. They were all filled with cargo, probably a year ahead of time, they were filled. So, they had to fly them in, and they brought them in Herc airplanes. They could take thirteen buckets in each Herc, and they'd fly into Nome and unload the thirteen. We had to get a hundred and thirty-six, because you had to have all of them to hook up the bucket line. It came out that there were three buckets that were left over, yet to come. By this time the big oil companies were after the Hercs, because the ice was breaking up in the far north, in Prudhoe Bay, and they wanted to get all the material they could flown in before break-up, and they had a lot of push. So, I found out what the value of an expediter in the North Country was. We hired a fella in Seattle who called me and said, "We're going to put a bucket on the regular passenger service airplane, on next Wednesday. Three days later there will be another one and then the third one." I told him he was crazy, but he did it.

I left Nome at the end of the dredge season in 1976, and hired out at the Ajax for Texasgulf. They had come in to partner with Golden Cycle, to rehab the Ajax and start mining gold once more. I was in charge of the Ajax, the bulk of it we had probably almost as many geologists as we did miners. The function of the mining crew was to keep the cages running, make diamond drill stations, and things like that, that the geologists needed for their diamond drilling, so they could test out different areas of the mine.

Most of this, just about all of it, was conducted on the Carlton Tunnel level, or the 3100 level of the Ajax. They did a lot of drilling, some of it, the fella in charge of the geologists at that time and the overall project was Harry Gair, who was a New Zealander. They had a bunch of pretty high powered geologists here. I've always gotten a kick out of them, because it was wet in the shaft, people wore slickers; yellow suits, pants and jacket. And some of the geologists who thought they were pretty good because they had Ph.D.s, they would put "Doctor Gair" or "Doctor Blah-blah" and so pretty soon, I noticed all the hands in the mine, they became doctors. 'Doctor Nothaus.' Oh, it went on and on. I don't know that the Ph.D.s ever got the message.

But, we did go ahead, we didn't have too much success underground. Mined a little, did a lot of drilling. That was the main effort. The geologists had samplers who did a lot of sampling, trying to see if they could find things that the old-timers had left. But we're still talking about milling the rock, and you had to have a higher grade ore for milling, conventional milling, than you would for heap leaching. And they shut down. We did, prior to shutting down, drive a decline, an incline shaft from the surface, into the Cresson Mine.

A set of crews ran over there, from the surface of the Cresson. They changed at the Ajax, but then they would ride over to the Cresson and they started that decline, as they called it, just a couple hundred feet from the Cresson shaft. It went down on the slope, and we had gotten two little rubber tired, they called them trucks. They were left over from the Uranium country, and they had a tricycle wheel set-up that had two little tires in the front which were right side by side together, and then the truck got bigger as it went back and had regular tires in the backend. They were called Young's Buggies, and the reason is because they were made by Young's machine shop in Monticello, Utah. We got a couple used ones. Golden Cycle, before Texasgulf came in, had decided they were going to go to the bottom of the Ajax shaft. They had had a contractor come in and sink the shaft to the 3350 level of the Ajax. They decided to use a Jumbo, instead of using jack-legs they would have these heavier drills mounted on big arms that would go out and they could be moved by compressed air, so it would take a lot of work away from the miner, and they could drill deep holes and break a lot of ground.

Well, the trouble was, the veins here are often kind of wiggly, and they'll shift around and they'll turn here and pinch there, so it's hard to go straight. You almost have to go round by round, maybe you have to move to the left on the next one, or hook it to the right on the next one. So they never did really use it. And we cobbed that together with an old drill carrier, a jumbo drill carrier that one of the used dealers up in Golden had, that put this new set of shells, they call them, the arms that hold the machines, and it was mounted on rubber tires. We used that over at the Cresson decline to drill.

Jumbo drills, this is a drill carriage where you have either two or three drills on these arms, and the miner can sit back at the cab and with levers move the machine around and do all the functions of running a drill, and then done hydraulically, which is much, much more efficient. But, we were kind of stuck with this jerry-rigged setup we had. We drove down 610 feet on the incline and then went out on the level. Couldn't find the ore we were supposed to be finding, so we drove a little working up to the diamond drill hole that had been put in by the geologists from the surface, and hit the drill hole and there was some gold there. Just what the drill hole had said it was. But it turned out that there were different formations of rock in the Cresson area, and in one of rock horizons the mineralization was gold mineralization. There was the structure of the vein, and it went down into the next rock type, but it was not conducive to mineralization.

So there wasn't any gold. We had a very little bit, as opposed to what had been forecast to work with. We thought that from what we had seen from the drill holes that we could go up and establish a little drift parallel to the main drift that we were in. With some drill stations in it, and we could ring drill, and shoot it, and have a pretty wide, I think it was about 18 to 19 feet wide, we could shoot a layer of that down. Then we'd go up a little higher, put a little drift in, set up a drill and ring drill it, and shoot down into the void where we'd blasted the rock previously. It wasn't quite as wide when we got up there. We went up to the third level, and you could touch with your shoulders the foot and hanging walls. The boundaries of the structure that had the gold in it just pinched off to zilch. So, things were not going too well.

Texasgulf shut the operation here down. When they shut down one time, no this was prior to driving the Cresson Decline and actually stoping in the Ajax, they shut it down because things, it wasn't a lot of production, and there wasn't any money coming in to Texasgulf. It was all going out.

I was sent over to a coal mine, as Manager, over on the Western Slope. We had a small open pit over there, right on the edge of the Grand Mesa. This was a surface operation, and we had to cut down through, there were four beds of coal, parallel to each other. The top three weren't too much, there was a lot of ash in it. So we had to excavate that through the glacial debris that had cut off most of the top three beds of coal, to get down to the twelve-foot high coal that was at the bottom. We had a high wall like you'd have in an open pit here, a hundred feet high, all of glacial debris. No coal or anything in it, and we got down to that coal and it had been mined out before. There were no maps that showed it, and so in coal mines you mine by the room and pillar method. It ends up like a checkerboard, sort of. We figured well, we'll mine out the pillars of coal that are left in the rooms that they took out. That was very hard because it got very dirty and contaminated from rock coming in.

Somewhere in here I was gone for a couple of months for health reasons, and when I went back they were still struggling with this coal, and then Texasgulf decided to come back into the Ajax. So they asked me if I wanted to go back, and that took about .02 seconds. So, we came back here. That's when we drove the Cresson Decline, and drove some drifts on the basis of the drill holes at the Ajax, where they thought they might have some gold ore worth hoisting. That kind of petered out, and cut way back, and I was 'retired'. Then I went to work for Silver State's operation. The years are pretty murky as NERCO took over Silver State's operation, and I spent most of 1988 as a NERCO consultant, setting up underground contracts at the Con Mine in Yellowknife, NWT, Canada. Was back here at the end of that year, and NERCO bought out Silver State. I went to work for them as an Engineer. Your Mom was working up there at that time, Sylece. Then I was back at the Ajax, and Hecla came in. I sure enjoyed visits with Norm Graham, the fella that had come down from Coeur d'Alene for Hecla. Talking about some of his shaft sinking episodes.

So, I ended up as Project Manager at the Carlton Mill, under my boss, Jim Muntzert. When he arrived as Manger, it was no longer a 'project', but an 'operation.' I became the Environmental Manager, I think that's what it's called, at least in polite society. By that time they had started some leaching. They had a small heap, and they were using a technique that Texasgulf had started, that Don Gulickson had set up for them. They put big tanks, they looked like big fire extinguishers almost, maybe twelve feet high and three feet, three and a half feet in diameter, out at the leach pad. The solution would drip through the leach pad, and they used wobblers, the sprinklers, they called them wobblers, because instead of being like a yard sprinkler, the head was loose, on an eccentric, and it would give you a different pattern. So it wasn't always the same pattern. That solution would go down through the heap and then they would pump from the bottom of the heap, through these tanks of carbon, made from coconut shells. Same stuff they use today, and that would become loaded with the gold. Then they'd take that tank, haul it back on a special truck to the Carlton Mill, and at the Mill they would put a hot solution through that and take the gold back off the carbon. Then put it through a little electro winning setup and the gold would be precipitated, formed onto the steel wool. It would then be collected and put through a vacuum press to get rid of excess solution, and then put in a crucible and heated, with a couple chemicals with it. Hopefully pour out gold, if you've heated it to about 2100 degrees Fahrenheit.

It was just a slight variation of what some people had been using. The Bureau of Mines in the early sixties, maybe late fifties, had brought up this leaching business and had advocated it. They did a lot of pioneer work, and so people had a lot of different variations of it. One fella even wanted to buy a fleet of cement trucks and put his ore in the cement trucks, and then it would rotate around in a cyanide solution.

Tell me about the first time you were underground?

I guess the first time I went underground, when I was in school in Golden, we had to put six weeks in, working at the School of Mines. They had a mine in Idaho Springs, and we drilled and blasted and mucked and did different things. The year prior to that we had another summer course, and had to do surveying, some of which was surface and some was underground. They picked probably the most miserable place in the mine for a survey. It was in a raise that connected to the Argo Tunnel. You may have seen the Argo Mill up in Idaho Springs. Well it was a tunnel, I think it goes back five miles under Central City, and ... I don't know that I thought much about it.... but pretty neat. I had never been underground before I went to Golden. I did, in addition to the survey and the mine practice course that we had to go through, I worked on some weekends. Went with a fella who was doing some examination and evaluation work of gold mines in the Idaho Springs area for pyrite. It was at the time when the country was very short of sulfur and they were actively looking into areas that had large deposits of iron pyrite, because they could get the sulfur out of that, economically. Depending on the deposit.

So we went in a lot of old mines where I'm sure we shouldn't have been. He'd core drill and all that, and made estimates of things; took a lot of samples. But the work we did in Idaho Springs, and different places, Caribou, and Central City, came to a screeching halt because they hit one of the big sulfur domes just off the coast of Texas and Louisiana. They had all the sulfur they'd ever want. So, we didn't worry about iron pyrite after that.

What kept you wanting to go back underground?

I don't know what it is. But like so many people, it's just something you'd rather do. But the opportunity did not present itself, and we work at what puts beans on the table. That's all. We hope that someday this operation will find enough material, gold, that will warrant an underground operation. It would not be like the previous ones. The economics of the methods, even when Texasgulf came back in, just were not sufficient, even with gold off the $35 cap, to be profitable at all. Some of the problems, of course, were the shaft at the Ajax. Small. When Hecla came in, they increased the size of the skips slightly, just so they could get a little more. One and a quarter tons, instead of one ton. Which over twenty-four hours makes some difference.

The Ajax shaft, never did figure it out, but there's a flaw down there somewhere. We had some skips get hung-up in there, and that was a real tiger to get them unstuck. And all that slows things down. You had to lower the crew and all their supplies in the same shaft that you wanted to hoist rock, which cut down considerably the rock hoisted. You had to allow some time for maintenance. It just was too darned costly for the amount of rock you could get out, particularly the amount of rock of the grade that was there, because you're talking maybe a quarter of an ounce of gold per ton, and that just doesn't really cut it. If you could have a not undiminished, but continuous supply, you probably could make a milling operation. Break even if you were lucky. Probably today even, you wouldn't do that. But the heap leaching advanced, it became larger as they used larger equipment. Surface mining is inherently safer than underground, not that you can't have accidents, but you're in a much safer environment.

Were there accidents at the different mines where you?

Unfortunately yes. Most of them were like statistics call for in underground mines. A fall of ground, a slab coming down or something like that. When I was at US Smelting, up at Lark, Utah, we had a couple of fatalities. One was a lessor, and we didn't have many lessors at that time, but on one of the upper levels, where the company didn't mine, and I don't know his name. He was a hell of a good miner. He was moving some cars back under a chute to load some rock to ship out and caught his head against the chute. That was pretty tough. We had another fella that, driving the raise, they were all ready to start drilling and they were taking a smoke, they hadn't checked the back too well and the rock came down. The fella was just sitting there.

Then we had a guy who was a Safety Engineer up there who used to fill in for the Foreman, Shift bosses and that. He was way up high in an old part of the mine where the company was mining, supposedly new rock. But it was all fill, from maybe ninety years before, maybe a hundred; we'd found old timber in it. He was checking the face to see if it should go for ore or waste, because we had to make that decision every day, and he didn't move fast enough, and the face came in. (chuckle) So I spent about three months in the hospital. Not too swift, to be the Safety Engineer.

At one point in time I had left Lark and went to work in Silverton. We drove the American Tunnel. It was an old tunnel, a mile in, and it ended up about probably 8x9'. We drove the second mile and started the raises to go up under the old mine. And that was the one that, many years later, and different companies, did a lot of mining there in Silverton and went up under Lake Como. Then one Sunday, when no one, the only shift of the week when no one is in the mine, Lake Como broke through. They said there were ten by ten timbers coming out the portal. We had just started raising up into the old workings, and the old workings, at the very tippy-top of them is where the high grade was, and I'm sure there's a lot of high grade came out of there. They were cognizant of the fact that the lake was right over them, and they had surveyors out there checking the depth of the lake, so they'd know if they were getting pretty close to it. What they did not take into account was that there had been some mills in that area and they had dumped their tailings in that lake, so the bottom was covered with tailings. Pretty slushy. It was by the Grace of God that it went when it did, with no one in the mine.

There was some high-grading down on the tunnel level. Hands took beautiful fluorite crystals out of there. Big chunks of them. Oh, they ended up in all the rock shops. One of the problems, they were beautiful light green, but if they are left out in the sun they faded to zilch. I mean they're still nice crystals, um, we kept looking for rhodochrosite, and found a little bit of it, and according to the consulting geologists where we had the rhodochrosite, we should have found some free gold, but to my knowledge we never found it.

This was in the American Tunnel, going in under the old workings. We used a slusher train there, that didn't work out too efficiently. Here, and most other operations, you would fill up a car from the face with your overshot loader. Then you'd have to pull that car back to a switch or to someplace where you'd stashed an empty car, and either have it come down from a little hoist, or from the side, and put the car in front of the locomotive and shove it into the face, so it could be loaded.

Well, to prevent that switching time, or save it, we had a set up where we'd have five or six cars, five ton cars, and they had rails welded to the top of them. The front car was cut way down, so it was low. We had a mucking machine that was on tracks so you didn't need rails. It would go into the muck pile and then go back to the car that was low, and it could dump into it. Then another fellow's back at the end of the cars, and he had a slusher, or scraper, and would bring that up to the front and pull the rock back and fill the cars, as they filled them then the muck would go back to the next car, and the next car. To save some of that switching time. Didn't work out too well. But, that's the way it goes in some places.

Was there high-grading at the Ajax mine?

We had high-grading going on here at the Ajax when we were working. Some people had a pretty good concept of where it was. And I'm sure we lost a lot of time. There's one or two people I know of, one of them let go, because he was just useless. I shouldn't say that, but he would have not been able to fill the poorest miner's shoes, in my book.

Anyway, there was certainly some high-grading. Not to the extent that it was when this camp first opened. That got pretty serious. One of the Judges said that the high grading was not an offense because the gold they took was a piece of real estate, and therefore it didn't matter. But it got so serious that eventually they put in change rooms where you had to hang up your diggers, and walk over to your locker upstairs to your outside-go-home clothes. They had somebody who must have had a weird sense of life or something, would stand there and watch the people go by and make sure nobody was hiding high-grade under their arms or legs. Then you had to have your lunch bucket checked, because that was very convenient.

When they got into some really high-grade stuff early on, you read tales of people who took it and broke it up into as much of a powder as they could, underground, the gold itself, and then put some of the rock drill oil on their hair, and put the gold in their hair. That was one of the ways. People had in their shirts and trousers pockets sewn in the sleeves and legs so you could stash gold in there. You've probably read tales about the people, that the top man would often know if a guy was coming up with some high-grade and he'd go over to kind of help him, because when he stepped out, say of the skip, and was climbing over, he might kind of tip over. Well, if he tipped over, why, that was kind of a clue to the boss.

One of the problems was the assayers. They had a whole slew of assayers in town. Some of them very competent people, but others were just fences for getting high-grade. The miner would get about half the price the high-grade if he was lucky. They had certain bars they would leave their high-grade in. They'd get a chit or something like that, so they had some proof they'd turned it over. Then three or four days later they'd drift by, and they should be able to get some money back. This continued up through nineteen fourteen, fifteen, sixteen. I ended up with a copy of some notes that a guy had made who was evidentially a spy for the Companies, and he wrote a lot in long-hand, "Joe Smith was seen in the 'Blah Blah Bar,' which is known for accepting high-grade." They didn't come to any conclusions, but they reported which bars were susceptible to picking up high-grade, and which guys were the ones doing it.

Gold mines really didn't help the war effort much. They wanted the hands, the miners, either in the service or mining copper or lead or zinc. Bullets and things like that that were more beneficial for the war effort. So that really kind of deserted this District for a long time, until after the war. Then after the war, a lot of hands didn't come back for one reason or another. Some found better ways to make a living, and some, a large number of people from Teller County went in the service, and a bunch of them didn't make it. Al Bebee, who was the Vice-President in charge of mining for Golden Cycle in the late forties, after the war, had lost two sons. One in the European Theater of Operations and one in the South Pacific. But anyway, things were kind of tough, prices were going up, inflation, post war inflation, but a lot of people did do some mining.

Golden Cycle opened the Carlton Mill because they thought they could save the freight costs into Colorado Springs, plus make a little more modern processing set-up at the Carlton Mill than at the mill down in Colorado Springs. All you can see of that mill now, as you drive down Highway 24, is the stack. Then many of the real good miners went off to the western slope about that time, 1949, 1950, to mine uranium. Because they could make a lot more money there than they could here. We were still under a cap of thirty-five dollar gold, with inflation going on, costs were rising and by 1960 things had shut down completely. The Carlton Mill had shut down about 1959, and they went through what ore they had, and I think the final date was 1961. And then things were dormant until the price of gold started to go up again.

In 1972, the government took the price off the cap of gold and let it seek it's own price on the market, like wheat, or corn, or whatever. And it shot up, because at that time it was thirty-five dollars in the US. Canada had a special premium if it was newly mined gold, so miners got a little more for it. On the Hong Kong market you could get about ninety dollars an ounce for gold. So, when they released the cap you didn't have to sell it to the government anymore, then it went up some.

Then the Hunt Brothers decided they would corner the silver market and raise the price of silver, which proportionately shot up gold, so it got, what $800? Well, that cut out a lot of uses for gold and silver. And the Hunt brothers never read their history books very well, because other people had tried to corner the gold market and the silver market a couple times, and it never worked. So, the price came down.

Do you believe in Tommyknockers?

I will never doubt that they saved my life up in that old stope it that turned out we were re-mining. I was checking the face with a pick. I heard a noise, turned away to step up on a timber set next to me and with my instant reflexes,
excellent coordination, natural physical speed, etcetera, came in second to a few
tons of rock from the face. Seriously, if I hadn't been given the warning it would have been all she wrote. As a result I just mashed a leg - no contest.

In the end, Anglo came in and took over the Minorco thing. I never did really understand that, but they really increased the size of the operation and the potential. Because the potential was to become larger. So, I consulted for a couple of years. Then it just became easier to become a volunteer. If I didn't want to do something, 'I just didn't have time.' But, if you're a consultant and the company wants you to do something, then you better damn well do it, or you won't be there.

Since then I have been a volunteer on various things like trails, etcetera, that the Southern Teller County Focus Group works with the mine on. The reason the Focus Group was formed was that many people came to Anglo and wanted them to build an airport on the leach pad, or pay Victor resident's water bills. Everyone, a different request. The Group was formed to get a community consensus of what most folks wanted the company to do for them. After getting a professional survey group from Denver to run surveys, most people wanted access by trails on some of the mining property around Victor, and to learn more about their mining heritage. The American Eagles and four trails were made, thanks to CC & V [a joint venture between Anglo Gold and Golden Cycle] and a bunch of signs installed to explain about the mining of various structures along the trail. We know over 11,000 people have used the trails to date, and only a fraction of visitors sign in. This has been a big boost of folks coming into the District for activities other than gambling. The Focus group has also been a factor in saving some of the artifacts and structures related to mining, in addition to the trails.

Hull City Mine

Channel 9 News

"The first thing you need to know underground mining is that your partner and each one of you watches over each other's back. The most important thing."

~ Brian Marshall

Born and raised in Cripple Creek, Colorado. 9-11-55. During the Gold Rush, my grandmother and grandfather on my mother's side, came here. Their name was Walker.

What did you think the first time you were underground?

Well, I was with Al Goodwin, and we were just oh, more or less, just pals. I wasn't hired yet, just visiting. Until I took an interest looking over the different minerals; what can be explored underground. He turned me loose with Greg Pratt, who taught me the ropes. Drilling and running the mucking machine. Of the hazards. Well, basically the first thing you need to know underground mining is that your partner and each one of you watches over each other's back. The most important thing. And that's what I learned. And enjoyed.

I'm basically claustrophobic in every sense. I think when I was a kid I got stuck in a culvert. I barely remember that. But a drift or a tunnel, which you're driving underground, it's 7x8', and there's plenty of tunnels to have your freedom to move about. Now, when your batteries go weak on your light on your helmet, you're dead, or you're lookin' for your friend who still has a fresh light goin' so you can get outta there.

What level did you work at the Gold Bond?

Four hundred. On the first level, we had our explosives there. So we'd take the skip up and fix up the fuses, and haul all the powder down. But you didn't take explosives down with the blasting caps in the same trip. It was a safety plan. The second and third levels were just abandoned workings.

Just across the way from the Gold Bond, we were checkin' out a mine over the way. It seemed to be pretty rich in its day. I heard tales the ore was so rich, they washed the walls down with water and collected the gold. The Cardinal. Just across from the Gold Bond. The Gold Bond's just nine-tenths of a mile south of town; Cripple Creek. You just drop off into that draw, and there's the Gold Bond. The Cardinal Mine's right across the way.

Well, we started the drift that way. Something changed in plans. I really wanted to go that direction. The track is still there, the switch is still there. It just hasn't got done.

All the mining we did was pretty much just solid, solid rock. Until we went through a pegmatite dyke, just full of crystals. From solid granite to these crystals and rare earth minerals, tourmaline, like I said, these little square cubicles. It wasn't iron. It was, as Al Goodwin found in a book: "Columbian Niobium Tantalite." Interesting.

It was just almost like a vein structure. Ah, maybe two and a half foot wide. Didn't need to be cribbed up, or shored up. But there was some ground that we went through, and it turned real rocky. It wasn't like a vein structure. We had to bar all these rocks down and soon we realized we were gonna have to start cribbing it. One of these rocks that we barred down was the size of a Volkswagen. (laughter) And I'm not kiddin'.

A lot of those were workings, see, the law is now that you cannot do a stope up anywhere from 100 foot to the surface of the ground. And the old timers didn't go that rule. They just kept on mining up and it all caved in. That's why there's so many stopes caved in like that. Well, there's a tale of an old boy in Victor. He parked his car in his garage one night. Next morning goes out to find it, it's gone.

Who ran the hoist at the Gold Bond?

Gregg would sometimes run it, and his brother John. There was a time John was runnin' the hoist. Okay, I was coming up from underground, and I saw some timbers that I wanted to check out. So, to ring a stop is one bell. Pull the cord as the skip's comin' up. I pulled the cord, checked the timbers, then rang the bells for up. When I got out, John was so mad that he had to stop the skip abruptly, that he was ready to beat me up. I thought about grabbing a 2x4" and clobbering him, before he killed me. But, it all calmed down. Al got in the middle of it, and stopped it.

Did Al run out of money at the Gold Bond?

Prob'ly shut down because of the money. It all goes hand in hand. After that we went to Lake City, Colorado. Silver and gold.

Once upon a time I was interviewed at the Mary Nevin Mine. I was claimed to be a good hand on the mucking machine. A dangerous sport, mind ya. If you put your head in front of the bucket, it could take your head off. If it came off the track, if it was too narrow in the tunnel or drift, it would pin you up and break your bones. But I was interviewed, and they had me on Channel 9 News, runnin' this muckin' machine underground. And when I got over to Lake City, Colorado, when Al Goodwin was runnin' some operations over there, Old Man Jim heard that I was an excellent mucking operator, and that's what I did. Just muck the rock.

And that's all I did, was run the muckin' machine. Suited me just fine. Well, actually, it was kind of a tough job, too, because they'd send me some huge rocks that needed to be broken up. So I'd have to get off the mucking machine and bust 'em with a sledgehammer. (laughter)

Okay, that was the Gladiator Mine. The other mine was the Fleece Ilma. Right next to the Lake San Cristobal, and it is the largest naturally formed lake in Colorado.

This was the Fleece Ilma. Anyway, I was lead miner there, and it was nothin' but rehabilitating this mine. I mean goin' back into all this muck and mire, and cave-ins, and timbers that had to be taken out. This mud was

the color of blood. Because of iron. We had to be careful not to let the water escape too fast because it would run into Lake San Cristobal, which was just below us.
Well, it happened one day. We gave away a big ol' dam of this reddish mud. Sure enough, it got away from us and went down into the lake. Boy, the Health Department got on us really big time. They didn't shut us down. We just had to get a $50 an hour backhoe up there, and dig some trenches and what-not to capture whatever overflow might occur.

What did you do there as lead miner?

Ran a crew of two. We had one geologist that was workin' there. Just gettin' a little experience. But he had this bad attitude. He was all the time complaining. Well, one day I flat out told him he could quit with his bad attitude, or he was goin' to hit the bricks. Not that I really had any right to fire him. But I flat out told him. He quit complaining.

Rehabilitation of mines is a whole different story than just mining itself. Like I say, mucking out the old timbers and mud, and re-timbering the spaces. But a new mine that you're working in, is just going in and drilling a six foot round. You have a burn in the center, it's 3 holes, and depending on the hardness of the ground, which granite up here in Cripple Creek is sometimes harder in different areas, and also depending on what you're dealing with, if it's granite, or phonolite, or breccia, or many different things you run in to. Generally you stay underground to listen to those explosions, to make sure every one went off. Count them all. Because if you go back and somethin' didn't go off, you could drill into it, and it could explode in your face.

What was it like to mine at the Mary Nevin?

Well, only makin' $5.35 an hour, for lead miner, it wasn't too bad. (chuckle) But, I think it was the second or third level at the Mary Nevin, they had some workings up there that I went to visit, a time or two. There was this little bitty hole, barely my size even. Barely able to climb through with all my safety gear, my battery, up through this hole. Now, I'm talkin' claustrophobic, now. I don't like little, bitty spaces. It went up into a kind of large room. They had a slusher up there; an air compressor, cable driven bucket. Went back and forth, slushing the rock to a hole, down below, to load up the ore cars. That stope was a big room.

Did you play jokes underground?

No. Like I say, we all were looking at one another's backs and practical jokes were not called for at all. At least I didn't care for it, I mean. I'm not a big man, but if somebody pulled a joke on me, I'd whoop somebody over that. I had one problem with one guy. He just had to find the time, when he had to take a dump, he'd go underground, in one of the prell sacks. Prell is a whiteish material that comes in bags. It's blown in the holes after the dynamite's loaded. With air compressed blow gun, and a suction type gun. And prell is nitrocarbonitrate. It was invented from a fertilizer, and it wasn't discovered until a ship harbored in Texas was blown up from this fertilizer. Well, we had to tell him it was just stinkin' too bad, and I don't know where these little flies were comin' from that hung around that bag he was poopin' in. But it had to end. That was that. (laughter)

I also worked with electric blasting caps, over in the Lake City area. After the Gladiator and Fleece Ilma of Al Goodwin's closed down, I went to work for the Golden Wonder up there. And oh, was I making good money. $15 an hour, plus $500 bonus. All this rehabilitation and dangerous work... I was making goood money. First started out as a welder there for $13 an hour, but they put me underground when they found out I had some experience. Well, here I am working with Jim Young, Senior, and Keith Young, his son. They're doin' this stope. So I'm the powder monkey. I'm haulin' all this powder up this dangerous stope, the ladder work and walk ways. I'm goin' "Come on!" And electric explosives, they scared me. I worked with an Uncle that hauled explosives up to Bowman. Couldn't have a radio in the car, because it might set off these electric blasting caps. Well, that made me nervous, just workin' over at Lake City with these things, I mean all different color coded. They'd go off at different times. Anyway, I got that job done, but didn't like it.

Plus, they ran diesel powered equipment underground, at this Golden Wonder Mine. After I left that job, I coughed up black smoke for over a month. They didn't have proper scrubbers, which are exhaust fume machines.

Did you ever encounter bad air here in the District?

Yes. I hauled Charlie out of the Gold Bond, once. When the safety lamp went out, and he was collapsing. Got to the skip immediately, and got our butts out of underground.
Low barometer will bring up the carbonic gasses into the mines. As long as the ventilation system was workin' properly, we could work. Even though some gasses were comin' out. There was a time, down at the Gold Bond, where we were drilling some long holes. That means many steels were put together, to see what's out farthest. Instead of drilling a six foot round, we might drill in forty feet, to see what was happening. There was a low barometer, and we had our lantern there. There was gas coming out one of the holes that we'd just drilled. Sort of an interesting thing.

Tell me about high-grading?

When I was working it took a half ounce per ton to even make it, to even take it to the mill. Half ounce per ton. You barely seen a quarter, maybe. But I've seen some wonderful high-grade. I'm not mentionin' any names, but pure Sylvanite crystals. Huge. Came home in a lunchbox, and this man had me pan it out. (laughter) Pan 'em out. He worked at the Ajax mine, and brought home some wonderful high-grade.

Al didn't mind me bringin' up pretty rocks. He and I were best buds. He's from back east. But, he started off young here with his father at the School Section Mine. He had claims to the Gold Bond for many, many years. He ran into the Farr's from Uvalde, Texas, and they were promoting the mine. Probably worked about six months out of the year, while the money held out. They'd stake him and run out, then stake him and run out. They staked Al for years.

Do you believe in Tommyknockers?

Oh, absolutely. Well, actually there was a time. Um... I didn't actually quite do a complete job of barring down. Of peeling off the rock on the back. See, to a tunnel there's the ribs, right and left, there's a back, and there's the breast. What you call the face, which you drill into. But anyway, didn't quite finish barrin' down properly. I was drillin' away and a rock hit me in the helmet. It got my attention. Didn't hurt me, but I thought maybe Tommyknockers were lookin' after me. Do a better job of cleaning the area, so there's no accidents.

Then there was one time an accident happened at the Gold Bond Mine. I was working top hand, above. They sent up a shipment of dump material. A guy had flipped the safety chain over one of the scrolls. Rollers, they go up through guides, up the shaft. And he flipped a chain over that roller, and when it got to the top and dumped that bucket, the whole top of the headframe came crashing down while I was beside it. Al Goodwin was there, and he just about had a heart attack thinkin' I was out there dead, dead, dead, underneath the headframe. But I heard it crackling and crunching, and I took off. I was 150 feet away before it came down. That was one of those, well, accidents can happen.

Carlton Tunnel

Riding the Rim

"When we crawled up that manway, seen all that blood comin' down there, we knew what we was gonna find." ~ *Jim McGee, Sr.*

I was born in 1926. I'm 78 now. A lot of people don't believe it. My Granddad was a grocer. He had better than $20,000 on the books when he died, grubstaking miners. He didn't die until 1931. He was old.

After World War II, I went to work in some of these little cap 'n ball mines that was still operatin'. I worked in the Ajax in 1961-1962. That's a deep mine, 3000 feet. I worked the Mollie Kathleen when I was 19 years old.

Mining?

Yep.

Was that a low budget operation?

Well, it was. They all were then. Because all the big mines were just about closed down. And there was no mill. I helped build the Carlton Mill. I did whatever they needed. Welding, carpenter work, whatever. Then a lot of the mines opened. But, they didn't stay long, you know, because the price of gold kept the same as it was. It didn't keep up with the economy. $35 an ounce.

I was about sixteen, I guess, the first time I went underground. Up here at the Lucky 7. It's right across from the Mollie Kathleen, over in that draw over there, just above the C.O.D. I went to work for two leasers in there.

You know, back then if you didn't want to work at a mine, you just went to the mine next to it. Got another job. So, I worked in quite a few mines. The Isabella, the Mary Theresa. I was workin' with Jim Robinson. He got killed. We was goin' down the shaft at the Mary Theresa, and we was trying to see if we could open it up further down there.

We used to always ride on the rim of the ore bucket, you know. That night he said, "I'm tired. I'm gonna ride this bucket up." He sat down in it. I was standin' there on the rim, and Red McDonald from Cañon City, was hangin' on that rim. And a damn lining board come outta that shaft above there, and they're soakin' wet and really heavy. A 2x6", about like that, 20 feet long, come down. He was ridin' in the bucket. Hit him right there, went through there, broke his pelvis, broke his bladder. Jim said, "Somethin' hit me." He never lost consciousness; he was a tough son of a gun. They put, I don't know, 4 or 5 pints of blood through him up here at the hospital.

When that timber come down, it hit my hat. That damn board went right between us. Knocked Red's hat off. We didn't even have a lamp because it tore my lamp, it just tore that right off. They found it two days later, 800 feet below where we were at. But ol' Red never lost his grip, and I didn't either. And then we had to holler at the other guys to stop the skip, because we didn't have no bell where we was at. So they stopped. I was so loud they could'a heard me to Victor, probably. So we took that board, about twenty feet long, and we had to hold it against that cable so we could bring that bucket up. Because where it was it would'a caught, and killed him. Hell, all of us. We had to pull him clear up outta that shaft with that board in him.

What they do with them mine buckets, is you go up above the shaft, then they put boards across the shaft, then put the bucket down so Doc Denman could cut that board outta him. We cut that board off and he said, "We'll leave the board in until we get to the hospital." And that guy never lost consciousness, the whole time we sawed that board off of there. He died the next morning. You think that 'aint something.

Did you go back to work in the mines after that?

One day I went. To quit. That was the last day I worked in the mines. Deep enough. I figured if I'm gonna fall, I wanna fall on the ground.

They had a lot of accidents in these mines. Explosions. One of the worst things that could happen to ya was when you were drillin', and the guy before shot his round. And if they don't count the shots, if there's one in there that's settin' there with a cap in it, and you drill into it... Remember ol' Doc Yarbrough? That's why he had bad eyes. He drilled into a miss. It was just damn lucky it didn't take his eyes out. What happened was, I think, when the other shots went off it blew it all out, but the stick that had the cap in it went off. That shot rocks all over; he was peppered full of holes. He was a tough guy. I liked ol' Doc.

Doc Denman would go right down there in the mine. He'd put on a pair of diggers, put on a lamp, go right down in the hole, if somebody was hurt down there. Francis, we had to haul her dad. He was drillin' in a stope, and a big slab of rock come down and pinned his head right against there. Well, when we crawled up that manway, seen all that blood comin' down there, we knew what we was gonna find.

But you still worked in the mines after that?

Yep.

Were you scared?

Oh, you get scared a little bit. What I used to hate and it's illegal, you're not supposed to do it, is somebody workin' on a level by themselves. Up there in the Saunders Shaft, they had me muckin' up outta a damn old caved stope. You're in there by yourself. Let me tell you what, it's quiet when you're by yourself, if you aint working.

I was back there about 700 feet from the shaft, you can't even see light there. My damn lamp went out. The battery went dead on it. Well, if you panic, you're dead, because you'll run around there and fall in a hole. So I just dropped down on my knees with both hands on the rails, and headed out 'til I got to the shaft. Got the skip down there, went up, got me a new light, went back down to work. (chuckle) You don't know what dark is. Dark out here, 'aint dark. Your stars and moon. You get down in one of them holes, and let me tell you what. When that light goes out you think all kinds of stuff. You gonna fall in a hole?

I was trammin' over a place that there was a big winze. They put slabs, you know, cribbing over there. Well, when you go to load the car, it goes down like that. Had to go pretty fast to get up the other side. I had to crawl over that, and I thought, you fall off of there, buddy, you're dead. So I just put my gloves on, put my hands on those tracks. You can imagine all kinds of stuff when you're in there.

What did you think the first time you were underground?

I liked it. For the simple reason you could have a big blizzard up here, you go underground, it's warm. Always the same. The Ajax is really warm. You get down there on the 31. There was places where guys, you know, would bring half a lemon or somethin' in their bucket, where they'd spit the seeds out. And that fine powder from some of them rocks, there was little lemon trees growin' out of them. Now that's hard to believe, but it's true. It's warm. I'd say it's probably 60, 65 degrees down the Ajax, and the further down you go, the warmer. What was fun in the Ajax, lightning like we had the other day, it knocked everything out. Knocked the fingers off that hoist, you know, to bring you up. See, there's big fingers, and when they come together is when the power hits 'em, starts the electric motor. Well, if it knocks the fingers off of them, you 'aint got no power. They can let the skip down, but they can't bring it up. They write a note on the damn powder box, a big ol' piece of paper or somethin'. "You're gonna have to walk out the tunnel, we have no power."

See, that old tunnel down there goes to Chapman's. That tunnel comes clear into the bottom of the Ajax, and you walk out of there. Oh, it's a long 7 miles. And you're walking along there, and you know where the water, from the water runnin' through there all them years, you'll hit a soft spot and fall in a hole. You'll be walkin' and bloop, your foot's down there that deep. There's damn muskrats and stuff in there. Oh boy, your light will be out there and you'll see these little beady eyes lookin' at ya and think, what the hell's goin on here? And there's beaver in there, too. And muskrats; there's muskrats down there this long. But boy, that's an experience, walkin' out that tunnel. You're walkin' along there, pretty soon the guy in front of you will go bloop, and about disappear.

Can you really see the little light of the tunnel's end?

Not until you get way, way, way in there. You keep wonderin' "Am I on the right track, am I gonna get outta this place?" I don't know if you remember old Charlie Wilson, he had a ranch down there across from Chapman's. The Wilson Ranch. He used to walk that tunnel to come to work. He was a tough ol' boy. He'd just come right up the tunnel; be there when everybody else got to work. Somebody'd take his time card, punch it out for him. There was some tough guys here. I remember when everybody used to walk to work, they never drove.

Where there pipes on the hills to supply air for the mines?

Oh yeah. Most of it came from, either they had their own generators, you know, little gasoline deals to pump air, or a lot of it came from the Cresson Mine. They had some big, big compressors. Yeah, there was the guy on the white horse. I'll tell you what, you could ride that line. John Phipps was his name. Ol' John used to... you could always see him at the mines. He'd go up, ride that horse where the pipe line comes along there. That's where you used to see that pipe all around the hills here. That was the air line. He was down here every day of the week. But it's a whole different world of people; we're a whole bunch different.

What kind of pranks were pulled underground?

Oh, I've picked my dinner bucket up and tore the handle clean off of it. Nailed to the damn bench with forty pennies. Or you'd go up to change clothes in the change room, your damn clothes would all be tied in knots, or nailed to the wall. Oh yeah, guys used to have a lot of fun.

Every once in a while some wisecracker would scrape the powder out. Take that stuff they're wrapped in, you know that greasy paper. Stuff it full of sawdust, put a primer in it, and light it and throw it at somebody. You thought a stick of dynamite was gonna get ya. Oh yeah, that was good. It'd wake ya up.

The worst thing I hated underground was powder headaches. Oh. Ya know, after they'd shoot a round, ya go in there too soon, that powder and that smoke and stuff, that glycerin. And ya never touched your head underground, because you'd get that on your hands when you was muckin' because most guys didn't use gloves. You'd rub your head, you'd get the worst damn headaches you ever had in your life. Boy, that stuff smells bad.

Aren't you were supposed to wet the rock down to help with that?

Well, that was a myth. They was supposed to, but they never did. Everybody was in a hurry to get the round out. You know, you'd get a round in, and a round out, every day, or you didn't make any money.

How did you get down the Lucky 7?

Oh, they had a gas hoist. They didn't have a big skip like they got at the Mollie. They had a bucket, and you'd just hang on that baby, hang on that cable.

All the ore went up in the bucket during the day?

Yep, put that in there. The way they dumped that, see, looked like a "V" sheet of metal, it hung up there above the shaft. When they brought that bucket up there they'd bring it way up, then it'd come back down it had a deal on it, on the bottom on a chain, and it'd hook in that thing and then it'd dump. Pretty neat. Them old miners was inventive, and tough. Strong. I'll tell you what, no bigger than I am, I carried a six foot stull when it was green, up a ladder, for a hundred feet. See, and when you go to town and get into a fight, you'll hurt somebody. Bad. You'd have loved to live back then, you'd have had a ball. People were different, they aren't like they are now. One person gets in trouble, everybody'd be there to help. Now somebody gets in trouble, half these guys will sit back and laugh at you.

What were you taught to call the frame over the shaft?

Gallas frame.

How do you spell that?

Gallows.

But we don't say it that way?

Nope. Gallas. But they spell it the other way. Or headframe. That came probably after the '30s. You'd be surprised how many of these old miners back then, how they got their ore out of the hole when they started. With a windlass. 'Cause I remember a lot of times we'd go up on the hill and watch these guys when we was kids, 'cause a lot of these places were up and going.

You were still allowed to do that after your cousin was killed by falling in a shaft?

Oh yeah. Well, we did anyway. My mother had five kids to raise by herself, because my dad died when I was three years old. If we wasn't workin' somewhere, 'cause we had to work to help her, we'd be up on the hill watchin' these miners. Every once in a while they'd say, "Oh kid, ya wanna run this windlass?" Why man, them guys had muscles of iron. We couldn't even turn the damn things.

Some of these shafts up here are inclines. Instead of goin' straight down, when they went on those veins would go like this... the one that Jim got killed on at the Mary Theresa, part of that was on an incline. You're goin' on an incline, then go straight down. Have another gas hoist underground. Had the air piped out, you know, for exhaust.

What about the mine under the old school in Victor?

Let me tell you what, there's not a building in Victor that don't have some kind of a hole under it. The State finally decided it wasn't safe for kids to be goin' to school there because it was undermined. They was afraid that miners workin' in there would cave that in, and all the foundation was cracked.

When I was playin' at the Gold Bowl, used to play baseball. You know that picture I showed you of the Cresson that caved in? There was a hole that big from the Strong Mine in that ballpark. Right there behind Travis Allen's shop in Victor, was the Gold Coin Mine. I don't know how many damn old cars they've dumped down there. And they still never got it full.

See, when they go down, most mines drift out. There's holes all over that damn place. I've worked under the cemetery over there. Got there from the Ajax, the Bobtail Vein. I don't know if you've ever noticed there's a little mine with the motor of a chopped in half car? We worked there. Damn, I never can think of the name of that mine.

When did you work that?

Oh, me and Al worked in there 1950 something. Silver State Jeweler, the guy that had that. It was gassy. Real gassy.

What did you do about that?

Just don't work when it gets gassy. When it gets cloudy and the air pressure gets low, you have to watch that. Keep a light burnin' down there, it's a gas safety lamp. If that goes out, you get outta there.

The Chicago Tunnel, right over here, right here behind the house, you can go in there and come out at the Anchoria Leland, up on top of the hill. You can even get clear over there to the Mollie Kathleen, if you know how to get there. And that thing's gassy. Ol' Percy Stone and me was workin' in there, we was in there sampling for, oh, I can't remember the name of that company from New York that was gonna do some minin' around here. We was carrying sample sacks. And when you did it you'd get you a pole, put six or seven sacks on each end of it and carry it out. Sample sacks are heavy.

Boy, I was walkin' outta there and pretty soon I hear his damn head hit the track. I looked back, and he was layin' there. He was out cold. I was taller than him, see, he come up to about here, and that gas was laying in there just along the bottom. Up along the back it was clear. Well, I was above it. I knew what happened, because I'd been in that before. I just held my breath to reach down. I got him by the shirt collar and I stood back up and that's how I drug him outta there.

We was almost outta the tunnel, but I still had to drag him about 50 feet. We got outta there, and they took him and I both to the hospital, 'cause once I got outta there, I got enough of that gas, when I sat down on that wheelbarrow I passed out. Woke up in the hospital up there, on oxygen.

He would'a died if I hadn't got him outta there, but they caught him in time.

Tell me about Calaverite?

Oh, there's black, then there's silver. Most of the black was in the Cresson Mine. When they hit the big vug in there. A tram car is about the size of this table, one tram car was worth $6,000. I worked in that mine; that was on the Isabella. That's where they hit it. And that come outta a room no bigger than, oh, maybe about twice as long as my living room, and about twice as high.

The Isabella's back up here, by Altman. On top of the hill. Well, the Isabella was right down below it. Then the Saunders shaft was right below that, then the Wild Horse was back this way. And you know what's funny? All the mines on that side of the hill was warm, real warm. That Isabella, you'd go up there some mornings and have to chop ice out of the shaft. I don't know why, because it's on the north side of the hill or what. Strange.

How would you get the ice out?

Just let it fall down the shaft. What you do is you put the skip up, then you get in a bucket and go down. You ever try to do anything on a bucket that's on a string? It twirls around. One guy holds the bucket with his arm by gettin' a hold of the cribbing in the shaft and the other guy chops the ice. It was cold. But you had to eat, so you did whatever they wanted you to do.

Was safety important?

Not until, really until the 1940s and '50s. Then OSHA and workman's comp and all those things that they didn't have before. Didn't wanna be paying no damn compensation because you didn't follow safety rules. Otherwise than that, they took a lot of chances. A miner was as safe as he made himself. You could go in there and you could be an idiot. Get hurt real easy. We shouldn't have been doin' what we were doin' when Jimmy got killed. We shouldn't have been down that shaft without a bell cord on that bucket.

Would it have changed anything?

Oh, no, I think what's supposed to happin' is goin' to happen. Once that timber fell, that was it, and it'd had a whack at all of us. But it got him. That's unbelievable. It went off my hardhat...

I've heard Doc Denman wasn't one to save fingers?

Well, it all depends. I've seen him sew a finger back on and you know, if it was beyond repair he'd say, "We can't save that." He was a doctor that didn't believe in putting you through a lot of pain to maybe save something that wouldn't work. Old Mike had a finger that was pretty well damaged and ol' Doc saved it. Somebody cut the end off their thumb, and he sewed it back on. I can't remember who it was. He patched me up a few times, you bet ya. Some from mining. Some from gettin' in fights in the bars. If you had a cut on your eye across there, or a cut on your hand or somethin' and you was about half inebriated, he'd say, "You don't need no anesthetic, you've already got some." And he would sew you up. He was the kind of guy if you hollered, the more he'd hurt ya. But he was a good doctor. Oh, I liked ol' Doc.

One time at the Mollie Kathleen, somebody left an ore cart out on the main track and I was really comin' out of there with the other car, and oh, when I seen it coming, I leaned back on my car, it was full of muck. The damn car that was behind me, I knew was goin' to run over me. I jumped to the side, but it smashed all 8 of my fingers. In fact, when I got married my fingers were smashed, I couldn't use them too much. It took a long time to heal and you can still see the scars. Just right between them two tram cars.

Billy got killed up there. Did you ever see a chute in a mine? Well, they leave an old steel in there, you know, to shut the muck off. Somebody left it in there, but they left it stickin' out and he was comin' on that air-tram, it stuck it right in his head. Killed him deader than a mackerel. He bled to death before they could ever do anything for him.

You see a lot of things like that in a mine. That's what tickles me, you see some guy smash his finger and go "Ohhh!" Why, you'd work all day with that. I seen Jake one time broke his wrist down there on the 28 in the Ajax, drilling. Know what he did? He took black tape, electric tape, he taped that damn wrist up and he worked 'til the shift was over. That was a tough old guy.

And you know when the skip's goin' up you don't dare flap your arms around, 'cause you'll catch it in the timbers and loose that baby. We'd get 'em every once in a while, me and ol' Dewey. We was skippin' over there on the Ajax. We'd stand up there on the ears of the skip, because they have little things that stick out like that, about half-way out. He'd hand me a five gallon bucket of water and them guys would be standin' right there, waitin' for the skip to come down. Splash! Man, they'd get me in that skip. That damn Jake, one night he bit me on the back. Right through that rubber raincoat. I mean it bled. He had me, I couldn't move. You're packed in there, about ten guys in there. I didn't let him know it hurt me. I wouldn't let him know he'd hurt me if it'd killed me. Boy, I tell ya, them guys were ornery. You had to be, to work down there. Because you know, you get down in there all day long, and rocks falling all around you, and guys getting hurt all around you and stuff. You get a little different.

Anchoria Leland Hoist

The Mad Mucker

"You'll never muck bound us. And the old boy in charge said, 'Oh yeah we will.' They bet a gallon of whiskey on this." ~ Alfred E. "Gene" Leaf

My birthday is 12-12-51. I was born in Cripple Creek, Colorado, at the old hospital. My dad was a miner, here in Cripple Creek. He came here after WWI, because his sister was here. She'd been here because her husband was a miner, and they worked here. My dad came over from Sweden when he was 13 years old. He'd already had three years mining experience underground when he came here. He had to go to work in the iron mines, because his dad got sick and they needed somebody to make a livin' for the family. So Dad got chose. All the brothers and stuff, they all had to pitch in. Dad got to go to school 'til the 6th grade. That's about it.

He didn't want me to ever be a miner. He jumped up and down lots of times, tellin' me, "I don't want you to do that! Nothin' but a lot of hard work, and you don't get nothin' for it, and it's dangerous." That's what he told me. Oh yeah, he told me the truth. But you know something, I'll tell you this: Me and my dad became extremely good friends because of this. Because I learned what it took to do this; it's anything but easy. I figured my dad had been doin' this since the time he was 8, 9 years old. Underground, hard work, miserable conditions. Whatever you want. And then he got stuck doin' this all these years, and wow. When I became a miner, I found out it really was tough. So I gave him a lot of respect, and me and him became good friends. We could talk on the same level and just have fun, and go on about our business. And do whatever else in life we had to do. It was really incredible. I loved my dad. He was really somethin', man.

When was the first time you were underground?

Oh, let's see. The first time I was underground, I went underground at the Mollie Kathleen Mine, with my dad. On the mine tour. Yeah, I got to go down there with my dad. I was just a little kid. We went down and took the mine tour, 'cause Dad was guiding tours. So was my Uncle Bill, at that time. Some other guys were guidin' there: Virgil Wilkerson, Loren, Johnny & Melvin Anderson were runnin' hoist. Frank Ferguson was the 'young buck.' Not really young, excuse me, I was just kiddin', but he was younger. Let's see, Ray Primm was guidin' tours there. These guys were all old miners.

When yer a little kid, bein' underground don't do much for you, ya know. But then it was a lot of years later, and I went to college and came back. Dad and Mom were workin' at the Courthouse, janitorin', so I took that job for about a year. Then they started openin' up the mill. So, I went over there, went to work in Sample Prep. This was for Golden Cycle. Golden Cycle hadn't yet became Texasgulf slash Golden Cycle, or whatever.

So, anyway, I worked there for 4 years in the lab. Then they said, "Well, we're goin' to have to close the lab down; they don't have enough work for us. So, we're goin' to send you to the Ajax." I said, "Okay, fine." You know, I had a job. So I went to the Ajax and they put me on the muck crew on 31'. And before that we had to go take samples all around, underground. Yeah, me and Chip made it as far as the El Paso, Roosevelt Tunnel level, taking samples. It was awesome. Got to see a lot of different things under there. We tried getting to the Vindicator lateral, but we ended up with a blockage. Stope caved in, so we couldn't get through there. Made it to the Cresson and all through that area.

Have you heard people say that you can walk from one town to the other underground?

Yeah, if you knew the direction. I haven't known anybody to do it. But it is completely possible, because a lot of the mines are connected together. But then, you might have to go up 300 feet to get to the other place. See? It'd be a maze, but you could do it, absolutely.

Well, to give you an idea, off the Roosevelt Tunnel level, it starts down below the El Paso, and comes up to the El Paso. Then goes up through the

Mary McKinney, goes up through that valley there. Because they had a big "Y" there. Go up the left and keep goin' til ya get to the Mary McKinney. You go around there to the right, you end up at the Elkton Mine, which was across the road from the old Mill, where Elkton the town was. And you go up there, then you could go to the Cresson, Vindicator, or just about to the Vindicator, the Portlands, you could go to them. All up this one tunnel level. The Ajax included. Yeah, you could go clear through there.

It was an incredible, incredible thing to see. That was on the 20, was the Ajax. And I think on 21, they had the pump station. Now mind you, they had doors down there that were 12" thick concrete. Vault doors. They had three of them. This was below the Roosevelt. They had a pump station down there. They had some big pumps. Bigger than that dining room in there. To keep pumping it up, because it would get seepage, see. Roosevelt helped, but it wasn't the answer, you see. So they had another pump station set up. What happened was, that tunnel that went back through there, goes back into places like the Vindicator, and it went on the other side of that hill over there to the School Section. But see, at one time, those doors were closed. They were closed and locked. Golden Cycle was down and put in the tunnel and all that stuff. And these guys, they got in a big war.

Which guys?

Oh, it used to be, well, this goes back in your history. You had the Golden Cycle Corporation, and on the other side of the hill you had somebody else. Golden Cycle wanted all the property. Well, they wouldn't sell. So, they just shut the water doors and let it flood. They'd do that to ya.

That's pretty bad, isn't it? Here's another one for ya. My dad told me this story, and he knew the lady real well. She was in the hospital, and she was very, very sick. And she asked one of the big shots to pay the taxes on her mine. He said, "Oh yeah, we'll do that." He didn't expect her to make it back out of the hospital. So they didn't pay her taxes, and then they bought it for back taxes. She got out of the hospital, she came back and she sued them. Yeah, she sued them to get her mine back. This is the kind of stuff that used to go on. This is big time money, we're talking. It's just incredible.

They weren't real nice to all the old timers, either. I'll flat tell you that. They weren't real nice about payin' 'em, they weren't real nice about treatin' 'em good. They expected you to work extremely hard, and that's the way it was. At that period of time, this is you know, after WWI, before WWII, they could do that. And that was expected of you. If you were a miner, you were expected to drill, shoot and muck every day. And if you didn't get it done, "Why?"

What did you think when you first showed up at the Ajax on the mucking crew?

What did I think about that? Eh, it's a job; it's okay. 80 degrees year round; 100% humidity. Wasn't too bad. When you went up it was kinda like getting a real shock when it's -22 out, when you came up off of graveyard shift at seven in the morning. Or when you came up after night shift, eleven o'clock at night, the wind was blowin'. Yeah, it was a little bit of a trip. You wear these rubber coats. You know, like rain gear. Well, if you could make it from the shaft to where you put your light up without your coat freezin', you were doin' okay. The coats would get so stiff, you couldn't even move the dern thing. (laughter) That's very true. That's very true. And see, they always had a shower up there. In the 'dry,' it's called. There's a shower room, because after you've been down that hot, if you just wear your wet clothes home, you're going to catch pneumonia or get sick or something. They always had a dry there for ya. It was kind of a neat place, I didn't mind that too much.

The only thing I did mind was when they sent up them great, big boulders. Which you couldn't shoot on the Station, because they were sinking shaft at this period of time. They had a crew down there sinking shaft below you. This was at the Ajax. And anyway, what the deal was, they'd send the muck out, dump it on the 31 Station. We'd muck it up, tram it over and put it in the other side, and it'd go to the surface. But they'd, oh, they had a little bet goin' one time. It was a gallon of whiskey. It was between the bosses. The boss of the sink crew, and the boss of the Ajax. And he told them, our fearless leader at that time, told them, "Well, you'll never muck bound us". And the old boy in charge of the sinkin' crew said, "Oh yeah we will." They bet a gallon of whiskey on this.

This was on my crew. There's two guys down there, and these guys, what they'd done is they'd take and they'd drill 16' holes. Vertical. Puttin' in a brow round down on 33 hundred. Like where you come down a shaft, and you have to go out at a 45 to make it big enough so you'll have a Station there. So you'll have enough room to move stuff in and out. So, it's called a "brow round." What they done is drill for two days, on that brow round. 16' holes. Yeah, clear around. Then they had that all set up, and here they go. Okay, they shot the round. And the first thing they did was they picked up the biggest boulder they could possibly find, and got it in that skip. I mean they had to take and dump the skip, which is 4x4 feet, and the guy on 31 who was runnin' the hoist had to go over there, pry the rock out with a bar, and then we were supposed to get this picked up with a 12B mucking machine, right?

Well, we couldn't shoot this boulder, and it was 4 by about 4 feet. I mean this thing is huge. So we took the mucking machine and we moved it outta the road as best we could. Which means our step was right in the road. So, then they sent up four or five more of these, and we kept movin' 'em around until we didn't have no room to move them. And then they sent up the big one. Rotten son of a guns. Yeah, they did that. So we got muck bound that morning. Our fearless leader wasn't very happy with me and my partner, but that was okay. (laughter) It was all done in fun. I'll tell you what, that was funny. Anyway, down at the dry a little later on they divvied up the spoils on the sink crew. The sink crew gave all the miners that were on the muck crew a big drink, so it wasn't too bad after all.

How did you do the mucking?

A machine. A 12B mucking machine, at that point in time. Yeah, that was just a little machine that runs up and down on tracks. You have to be real careful with it because you're standing on one side of it, and the rocks go right by ya. The whole bucket tips backwards and throws this in a car behind you. So you have to be real careful.

Did you ever have accidents?

Me? No. Noooo, not me. I was too chicken to have a wreck. Well, the first time when I was runnin' the 12B, this was down on 33 after they got it down that far. I was behind the chute, runnin' the mucking machine. But I had about a quarter inch between the chute, and the handle, to run this machine. If you don't think that was hilarious. I'll tell you what, it spooked me more than once when the bucket would turn, because it can pinch you. Take your arm right off, or your hand, or whatever. So you have to be real careful doin' that. There's been a few times I've been a little spooked here and there. A lot spooked, not just a little, okay.

What do you think of Tommyknockers?

Tommyknockers? Ahhh, there such good little critters. We was back there, taking samples one time. We were at the Anaconda Tunnel, as a matter of fact. There was four of us headed back in there. And you have to go by this one chute, right? Anyway, we was all headed back in there and the first guy goes through. I was second. I heard some tappin', so I just went right on through. Other guy, he stops. It's a good thing he stops, because the chute gave way right there. About 3 tons of rock just dropped. 3 tons of rock is only about that tall. 4 or 5 feet tall. But the thing is, if ya have it hit ya, it's going to hurt ya.

You heard tapping?

Yeah, I heard tapping. I don't know what the other guys heard. But the other guy stopped, and I just kept going. And it fell right between us. It was just incredible. Do I believe in Tommyknockers? Yes, I do. Lots of times. That's just one instance.

There were other times you heard sounds?

Yeah, I heard sounds. Once in a while they would make ya move. But, that was the only time I ever remember hearing that very distinctly, and something bad really did happen. I remember that quite well.

Did you think in the moment it was Tommyknockers?

Yeah, I did, as a matter of fact. Now, another time I had a one ton slab of rock come off, it was on an incline of about a 45. I was standin' there against the wall. This rock slid, it gave me one scratch on my shoulder, and it landed right next to me. It came down about two and a half feet. Yeah, I didn't hear no tapping that time. But, the thing was, somebody'd yelled at me. We was in a stope. Somebody yelled at me and I just, you know, turned. If I had been over an inch or two, I'd have been dead. I wouldn't be telling you these stories. It would have caught my shoulder and pulled me under it.

Did you work on ice in the Mary Nevin shaft?

Oh yes. Me and Ronnie Myers. Lawrence was runnin' the hoist. We had to shut the mine down because the ice had formed on the guides, and you couldn't drop the skip through it. So, Brad said, "Okay, you two guys go down there and get that chopped so we can get on with it tonight." So we started out at about seven in the morning; started choppin' ice. We finished about 3:30 that afternoon.

Now, in order to chop ice in the shaft, you have to get out of the skip, after it stops on the ice you have to get out and try to get to a ladderway. My job was getting out of the skip, getting across the timber and chopping the ladderway, so I could go down and chop underneath the skip. The ladderway, you don't have rungs, it's just a solid sheet 'a ice. So, I chopped a hole in that and proceeded down, nice and careful. Hung on to the rung, after I found it, with my arm. Not with my hands, with my arm. I kept chopping away 'til you find another ladder rung, and go on down. Then you proceed to chop the guides, and Ron was in the skip, and he was choppin' ice on the wall plates and end plates. And we went up three times, I think, that day. Up to our hips in ice. Had to dump the skip out.

So we got down through the ice, and we got it all cleared up. And the next day, guess what? More ice. It went on for maybe two or three weeks. Then they got a heater, like a big vent fan except it had a heating unit in it. They blew air down the shaft, warmer air, and that kept the shaft clear. But there for a while, Man, it was pretty funny. (laughter)

Chopping ice was bad duty?

Now, you can look at this in two different ways. I want you to know it wasn't fun, because you got cold. But the danger, if you like danger... all miners like danger. They're all livin' on the edge. I couldn't do that anymore, I'm too old. But back when I was younger, that used to be just the "A-Big" thing to do. "Hey, let's see if we can get this chopped. Okay, cool, Man. I didn't slip and fall, whadda ya know. Isn't that neat."

Adventure?

Yeah. That's what mining is, it's a big adventure. You hit the nail on the head on that. It is. It's really going someplace that nobody's been before. And it's just like exploring, you're going "Wow, Man, what's next?" It's really cool. You don't take in to consideration all the dangers or problems. You just take into consideration, "Wow, what's next? That's cool." And that's what really, that's what makes people like it so much. I used to love it.

Me and Junior, well, a prime example of this. We were drivin' drift on 33, and we'd get to turn the corner, and we hit some hot rock. So we get up there about oh, another hundred, hundred and fifty feet, and anyway, we drove a round, shot it. It was solid red granite, okay? The next round had a big black ridge in this red granite. It came up, we don't know where from. But it came up, it was there, and it was black. I think it was breccia. But it came up and it went across, right in front of ya. If you're standin' there, it's about 4 foot high and it's like 'Wow!' You never know what you're goin' to hit next. You never have the slightest idea. It is cool. It is so much fun when that happens.

What do you know about Calaverite, and our local telluride ores?

Calaverite, let's see, there are four different types of ore related to that. Calaverite, Petzite, Krennerite and Sylvanite. Each of them contain a specific amount of gold. Like Calaverite is 80%. Sylvanite varies, because it has a higher content of silver sometimes. Sometimes it's 20% gold, and 60% silver. Other times it's 50% gold and 50% silver. It can vary. Petzite and Krennerite are rare, but they're found here. It goes down to things like tellurium, and the gold content's about 20 to 30%.

Is this District unusual, geologically?

Oh, very. Very complex area, within a complex area. Yeah. You know who taught me that? Frank Ferguson. Old Furgey. You can have all the geologists in the world come up here and they'd be lost. Because it's so varied. Every mine has it's own little tricks to it. Like over on, let's say, the Victor Mine. They have a talc seam over there that carries gold. The fluorite vein doesn't carry anything. Then at the Mollie Kathleen you have fluorite, and that carries white iron. The white iron runs about 3 ounces of gold per ton, in different places. What I'm trying to tell you, that same white iron will run nothin' in one six foot section. In the next six foot section it'll run 3 ounces a ton. I mean, you know, that's just some of the stuff that goes on. At the Ajax it's in the tellurium, which is gray colored. But that's right next to the red granite. And then there's a quartz slip in there, also. That's on 31. Now, in different areas, like the Cresson, it runs in your tellurium, but it runs in all sorts of different sizes.

There's a hole at the Cresson you could probably put half of Cripple Creek in. It goes from one end of where the Cresson shaft used to be, clear across the High Road, over on the other side. It's called 17-4 Stope. And I've been privileged enough to see this. There's not one stick of timber in it. There's not a tree long enough to do anything with it. You walk for twenty minutes at a very good clip, and you don't realize that you were underneath this. They left a ten foot cap over the drift where you're walking. You did not know that you were even on the side of this thing, until you get out to the end of it, or some of the end of it. And a big boulder came through, abut the size of a D-9 Caterpillar, and it broke through that ten foot of rock. The rock is still layin' down there. Where it came from, we had no idea, but no doubt about it.

I was privileged to see it, and I don't ever want to see it again. 'Cause, I'll tell you what, that is some of the most undermined area I have ever seen. There's things up there, you don't know what's gonna come loose next. They had bridges across it; they had all kinds of stuff from what Jack Hamm told me, and he done a lot of research in that area. He was one of the geologists for Texasgulf at one time. He did a lot of historical searching. We were back there takin' samples, so we just thought we'd take a trip back there and take a look at it. It's on 1700. You can take three mine lamps and a high powered flashlight and not see the other side. Total darkness. But I've seen the other side of this thing.

It was a huge, huge hole in the ground, is what it is. Anyway, there's all sorts of different sizes underneath the Cresson. You have 50 foot diameter holes that go 4 or 500 feet, straight up and down. Yeah. At one time, see, the Cresson did not make any money. They were about ready to shut it down. So this old boy they hired, he decided he'd make big, big holes instead of little holes, and that way they could keep the mine open, because it was all low-grade. So that's what they did. After they got down there, then it became one of the richest mines in the District.

What are some of your favorite mine stories?

Well, one of my favorites is there were three guys in this stope. I was one of them. Jimmy Jr. was one of them, and Rollison was the third one. We were all up there with 83 stopers, drilling. I'm in the middle, Junior's on one end, and Rollison's on the other. And we're all drilling away. We have these ear plugs in. I hear this strange thing, Man. I don't know what's goin' on. You know, you hear somethin'. So, I shut the machine down and went over. Junior, he's goin' along okay. I thought I heard somethin' else, so I went back and seen Rollison. Here's Rollison back there just singin' up a storm, I mean to the top of his lungs. He's just screamin' back there over the machine, "Come on Baby..." So, I went back and started drilling again, going "Oh my God, where did we get this guy?" (laughter) Some of the craziest stuff in the world. That's one of my favorite minin' stories.

Another one is, oh, me and Lawrence were down at the Mary Nevin, and the hoist broke down. So, Brad tells us we have to have it fixed by morning shift. This is 8 o'clock in the morning, right. So, we tear the hoist down, send the jack-shaft to Denver to get another one built, right? Then me and Lawrence were getting all the parts and everything cleaned up. "Oh, what's goin' on there?" The jack-shaft shows up about six. We'd already been there 14 hours. So anyway, me and Lawrence, we spend the rest of the night. No phones down there, so we couldn't call our wives. We told Brad to tell them, but you know, he went to the bar. So anyway, me and Lawrence are down there workin' away, and pretty soon here comes our wives. "What are you guys doin?" "Well, we're workin' we're trying to get this thing put back together." "When do you think you'll be home?" "Well, we don't know. Just go home, and we'll try to get there tonight sometime."

Well, anyway, they brought the cable in. We had to put new cable on and they didn't get a long enough piece. So we were out there playin' with the headframe, tryin' to get it to wrap right. We were there until 8 o'clock the next morning. We put the old cable back on and it wrapped just fine, so the boss says, "I think we'll just go with that. You guys go home and get some sleep. Be back tomorrow." "Okay, Brad." (laughter) Yeah, that was like a whole day. 26 hours of workin' with this other guy, and we just had a ball. I mean seriously, we had a lot of fun doin' this. But our wives, they didn't know what to think about this, they're goin' "What's the matter with these crazy guys?" (laughter)

Let's see, what was another favorite one? Oh, it's the time Jimmy, Senior.... He's a big man. Flash, he's like this tall, and a real skinny guy. So anyway, we're down at the Ajax, the old man's drillin' there on the Station. He said to Flash, "Hey, Flash. Come over here. I gotta go down to the tunnel, and you know... could you finish my hole for me?" What the old man had done, is he left the machine cob turned for him. And he can hold the machine down with one hand. It takes a lot of pressure. But Flash, he's a little guy, right? The old man done this just bein' ornery. Flash was little, and it takes a lot of power just to hold the machine down. So anyway, Flash gets up there and pretty soon he's goin' up, and up, and up. Pretty soon his head's against the back. The old man reaches up and shuts the machine off. Grinned at him and said, "I think you're a little light in the butt for this, Son." (laughter) I tell ya, there's always stuff like that goin' on.

Were there a lot of pranks?

Yeah, they used to do a few. You want to hear a good one? Okay, if this guy's really been givin' ya a bad time, what you do is go up and cut his airline in half, or cut it at the end on the machine, or where it goes into the airline, right. You cut that hose, you put a ping-pong ball in there, right? Then you put it back together. You turn on the machine, it goes "Trrrrr-blah blah." Stops. Then you let the air off, ping-pong ball goes back down, turn the machine back on, it comes back up and stops your machine. You have to cut your hose, take the ping-pong ball out, put everything back together, and you can drill. I only remember one guy they done that to in all the years I was here.

They didn't like him?

No, they didn't like him. You know, most of the stuff was, like that bet I told you about. And you know, you always had that goin'. Everybody had a bet goin'. Like "We're goin' to beat you guys!" "Oh yeah, let's see. We'll bet you a gallon of whiskey on it!" "Okay, fine." And this went on. Then everybody goes out of there rip snortin' drunk and then goes home.

Did people drink and go to work?

Sure. I worked with this one guy, I won't mention his name, but he spent more time at the bar than any one man I ever knew. He could drink a quart of whiskey, and you'd never even know the man had had a drink. He always had a good shift when he came in three-quarters of the way looped. Because he could out-drill anybody on the whole mine. But he'd have a few toddies and was just fine. Okay. He never was a safety hazard, he never bothered anybody. It was his own thing. He was a good man; he was a good miner. A real good miner. I have a lot of respect for the man.

If you did your job, they didn't care if you drank?

That's it. But if you became, you know, if you were too drunk to do your job they used to set you on the bench.

And there you'd stay?

And you better stay there, until you got a little bit sobered up, or else you might hurt yourself. And I'll tell you who that is, that's me. Because I came in too drunk to do my job. I was down at the bar havin' too many toddies and havin' too much fun. Forgot it was time to go to work. And I didn't call in sick, so I went to work, yeah, and they put me on the bench and says, "You just stay right there, and you'll be just fine."

Were you underground?

Yes. The bench is 3100 feet down, and off to the right. (laughter) Hey, I've been there. I can tell ya all about it. So, you know... drinking... well, in mining, drinking is part of the whole thing. It is. It's part of the mining game. Yes, it is. A big part of it. Because you're down there, you work extremely hard, you don't know if you're going to make it up the next day, or not. Or that shift. Or that you're not going to come out of there hurt, or dead. You don't know. And you know, it's just part of the game. You make it out of there, "Oh, time for a brewski. Okay!"

Oh, I got to tell you this story. The graveyard crew at the Ajax was goin' to go down to Zeke's and have breakfast. Well, Ohrt Yeager was runnin' Zeke's. He did not like to fix breakfast. Especially for a bunch of crazy miners, okay? And so we all load in the trucks and go down there. And Ohrt was cookin' him a trout in a little pan. Oh Man, he just about had it perfect, I'll tell ya. It looked really good. In walks all these miners, and we start orderin' breakfast. Ohrt got so damn mad that he took the pan and threw it in the trashcan, fish and all. We all busted a gut. He got so mad at us. Sounds just like a bunch of miners. Done this just to be ornery. You know, miners. It kind of gives you a feel for what miners are like. They're ornery little critters, always havin' a lot of fun.

Sometimes at others expense?

Well, you know, absolutely. But Ohrt always run a good bar, and everybody liked goin' there. I have a lot of respect for Ohrt Yeager, he was a good man. He used to carry guys drinkin' and carrying on; he used to carry 'em. Course now and then they'd take their whole check down there and give it to him. (laughter) But, that was Ohrt, you know. That was the minin' game. It was really fun.

Like a different world then?

Well, you could do that then. Now if you had a bunch of crazy undergrounders up here, with all their laws and all this stuff... my word, they'd have the SWAT Team out all the time. "What are these guys doin'?" Hey, I'm just bein' straight about it.

I mean like, you know, 4th of July in Cripple Creek. When I was a boy, there wasn't such a thing as fireworks up here. We had, well, this one individual got a case of powder and he went up here on Gold Hill. Now, for 4th of July, he set that whole case of powder off. Yeah, rattled every window in town; everybody thought it was just great. And the police, they cruised up there nice and slow. They weren't in no big rush, you know, they understood. This was 4th of July. They looked around, "Okay guys, well we can't see 'em. Don't know where he's at. Don't know who did it." Came back to town. Like hey, that's okay, 'cause nobody got hurt, it was just fun. But now if you do that, well, you'd be in the league of oh, I don't know, Al Capone, (laughter) or somebody like that. Yeah, you'd be in bad company. I remember that quite well, because the house here was not fixed up. The windows about rattled out of the place, because the putty was weak. The windows dropped about an inch. My dad went out there and says, "Ew, I guess I better putty these windows back in, it took it good." I'll never forget that.

Well, anyway, let's see, another of my favorite mining stories was, oh, the Anderson brothers. Two old-time miners. They were workin' on the School Section, tryin' to find the shaft. And anyway, I went up there to see 'em, 'cause I liked 'em. I wasn't minin' at the time, I just went up to see 'em. And here's John and Melvin down in the hole. Had the bucket in there and they're standin' there arguing about a quarter of an inch. "No, John, it's over here." "No, Melvin, it's not. I want you to know, it's over here." "No, John, it's over here." "Now Melvin, don't loose your temper. It's over here." This is all over a quarter of an inch. And I'll tell you what, I never laughed so hard in all my life. I thought I was gonna die laughin'. 'Cause here's these two grown men standin' there fightin' over a quarter inch of where to put the timber. That's all. A quarter inch on one side, or a quarter inch on the other. I stood there goin, "Oh my God."

A friend of mine was up there with 'em, and he was runnin' the bucket truck. And I says, "Does this go on all the time?" He says, "Oh, it's just constant." I says, "How do they ever get anything done?" He says, "They don't."

Anderson brothers were a couple of real top notch miners in the town, here. Their dad was a timberman. John and Melvin Anderson were good miners. John and Melvin, I had a lot of respect for; they were really good people. And I want you to know, Melvin lost his leg in California in a mine accident. He lost his leg from the knee down, and he had a prosthetic. And you couldn't tell it. That man could walk timber. He could walk timber in the shaft just like you or me would. Now he was a little more on the ornery side, but he was still a real good miner. Johnny, he was more of a hoistman. He done the hoisting, and you know, the other stuff. He was an excellent, excellent hand. They both were really good miners. They really knew their cookies, I'll tell you that. But, just to watch them work together, Man, it was a riot. They wanted it absolutely and positively correct.

Were there accidents in mines where you worked?

Oh, yes. We were goin' in on graveyard shift and they just got down to 33 hundred, at the Ajax. A guy by the name of Jim Sandborn was runnin' the slusher. A guy by the name of Freddie Shreck was runnin' the cryderman. The slusher pulls the rock towards ya. It's just a bucket; runs off cables.

The Cryderman, it's like a shovel, okay? But it goes straight up and down, and it'll swing, like this. It has hydraulic arms on the bucket. You pull that up and then you pull it over and drop it in the skip, see. Then you have to put the guides back in, then they take that bucket to the surface. But this machine sits up about 30 feet in the shaft. And Freddie was up there runnin' this and Jim Sandborn was down on the station, runnin' the slusher. Well, the cable snapped on the slusher. So, as the cable came around Jim, it pulled him into this slusher and gave it more power. He couldn't shut the machine off. This cable's wrapped around him and it's startin' to pull. And it started cutting. He had cuts one inch by one inch, from here clear around. Wouldn't 'a been for Freddie, who slid one hanging rod they had set up. A hanging rod is what holds the shaft together, it's a little half inches piece of steel, round, that's a hanging rod. He slid one for 30 feet, to get down there and shut that air off before it pulled him in half.

So, we come in on graveyard shift, 31 hundred is covered in blood, and we asked what happened. And they told us that's what happened. They took him to the hospital. Jim did recover. He is now a Colorado State Mining Inspector. He's a good man. He's a real good inspector. But ya know somethin', he won't be able to do what he used to, because of all the surgery he had to have. It took him just about a year to come back. He got hurt real bad. It wasn't even a bit funny. But he managed to come out of that alive, and he's a good inspector.

Then another guy I seen get hurt was oh, tryin' to remember his name. Oh well, what the deal was, is ridin' the train, underground muck train, was goin' down to 3-Mile Station, in the Carlton Tunnel. Well, anyway, this guy stood up and he got caught between the motor and the wall. It rolled him three times. Then he fell in between the tram car and the motor. So, we went and got the Stokes stretcher and the backboard and all this stuff. Ran it down there, put him on it, and sent him out. He ended up with both busted shoulders, two or three dislocated vertebrae in his back, and a busted hip. He was a pretty tough old boy. But anyway, he made it out of there okay, too.

But then, a couple of other guys didn't. I'll tell you about that. Paul and a guy by the name of P.J., he was just hired. This would'a been on my shift, at the Ajax, except I got fired about three days before this incident took place.

Anyway, they had a 3 day weekend and somebody shut off the fans to 33 hundred. And these guys came in on graveyard shift, which was the first shift back. My ex-mucking partner, Gary was there. And what happened was they had a shaft wreck down below 31, and they had timber out. So they were gonna go fix this. So they go down there, but they didn't know the fan was off. And the gas came in. They dropped below 3100 and the gas got them. At the Ajax.

Somebody didn't turn on the fan?

Well, it got turned off. I don't know who turned it off. Either an accident or something happened. Just one of them things they should have been watching more closely, but they weren't. And the gas got them. One guy fell down about 200 feet. Another guy just collapsed in the skip. And Gary, who was down there... the hoistman couldn't move the skip. Nobody could get down there, because the other side was blocked off. So, Gary was down there but he couldn't do nothin'. The hoistman kept flashing the lights and carrying on, but nobody was there. Nobody could do a thing.

Why couldn't Gary do anything?

He couldn't move that skip. He didn't know what they were doin' down there. Gary was on 31, these guys were below 31. They had the skip down there. See, if you cannot ring that bell... the hoistman didn't get no bells. He got zero bells. They gave him a stop, and that's it. You are not, by Federal Law, allowed to move that hoist. Period. Nobody can ring that bell but the guy who has it. So, they sent a team down through the Carlton Tunnel. It's six miles long. Into 3100 of the Ajax, and they detected the gas, so they turned the fan on. It was too late for them guys, they're already dead. And Gary and another guy just about got killed because the gas came up on the Station, up to about shoulder height. They didn't get gassed; they ran. They ran down to the Carlton Tunnel, where there's a fresh air source. Gas will dissipate there.

They had to wait until the gas got cleared. Then they had to go down and see what happened to them guys. They got to the fresh air source, cleaned up everything. You know, got all the gas out, checked it with a lamp, then they climbed down the ladderway. That's what they found. That was pretty bad. They were dead. I wasn't there. I went up to the boss a week later and told him, "Hey, that's the best thing you ever did was firin' me. Thank you." (laughter) And walked off.

Did you work at the Ajax again after that?

Oh yeah. Sure. Hey, ain't nothing like the old days, but you could usually get a job at the mine.

Did you muck for a quarter a car?

Oh now, that was at the Mary Nevin. And what the deal was, see, they gave ya so much per car to tram this stuff. To get the car from the chute to the skip. They paid ya so much per car. Yeah. That's called Contract. This was for fifty-cents a car and it was somethin' like seven bucks an hour. Wasn't very much.

That fifty-cents counted a lot then?

Oh yeah! Oh, Yeah! That was big time. I was back there pullin' chutes tryin' to make money so I could support my family. That's what that was called. Anyway, I think the best I ever did was 60 cars in one day. Yeah. Course, I aint goin' to tell ya how I got that, either.

Oh, I'll say. I had some friends back there. I'd been workin', you know, trammin' cars for, oh, about a year. They knew I was givin' it heck every day. So, one day they climbed down, and they said, "We're goin' to give you a hand." So they got back there and helped me. Then I kept pushin' them cars outta there as fast as I could go. (laughter) Hey, real nice folks. Real nice guys to do that for me. I didn't expect it. I never asked. But that's the way miners are. They see you're tryin' and they know that you're not makin' a lot of money. They're the guys that are makin' the money, because they're drillin' and shootin' and stuff. Trammers don't get nothin'. They knew that, so they came down and helped me. I made all kinds of money that time around. It wouldn't be nothin' now, but it was great then.

See, Brad, he was the guy in charge of the mine. I always beat the nightshift. Hands down, every week, my paycheck was always bigger than theirs, and there was two of them. See, they were gettin' a quarter a piece. And I'm the only one down there on dayshift and I got fifty cents a car. I used to beat them hands down on number of cars, every week. Then when those guys helped me, I really beat 'em that time. Brad came to me, "How did you do that?" "Well, just got lucky, Brad." (laughter) I never told him.

In order to get them cars, sometimes you had to shoot the chute. You'd get a hang up, see. And now, here's the tricky part. You have a ten foot pole, piece of wood with a wedge on top of it, and three sticks of dynamite on it. You have this one chute that you have to lay down in it, and work yourself back up, backwards, and then put this stick up against the hang-up. Oh, we used to love to do that. Believe it or not, sounds crazy. "Oh the chute's hung up. I'll fix that!" Okay, send me down powder and primer and fuse and all this stuff. I'd go back there and fill all this stuff up, get back there and put it in. Put the boards back in. Then: "Fire in the Hole!" Boom! "Well, what happened?" "Well, nothin'." "You guys do it again." "Well, okay!"

That can be dangerous, freeing up those chutes?

Yeah. It can be deadly; it can be very deadly. You can get crushed. If you're in the way the only thing you can do is... I had it figured out though. What you can do is you can slide out over the car.

How would you do that?

You just give yourself a big push, and you go right over the car. Try and get on the other side before it hits the car, tips it over, and buries the whole damn drift. And you get outta the way. That's the only way you're gonna do it. Well, that's one story.

Another story I have that happened at the Ajax, we had a sub-level and then an upper level. We had guys on the sub-level, and we had guys on the upper level. Pullin' in rounds for a stope. Well, me and Junior were on the top level. The guys on the sub-level were shootin' boulders. It's all done with electric caps, okay? So anyway, me and Junior have two rounds sittin' up there, sixty holes a piece. Filled with dynamite and all set up, ready to go. So anyway, what happened was, we went down and we shot it, right? Two rounds up above did not go. Electricity followed the line of less resistance, and shot the boulders. So anyway, me and Junior wanted to stay there and shoot it that night. They said, "Oh, no, you guys go home."

So next day, our boss sent me up there to check 'em out. Meantime, he had Jaccard put a charge in the chute that I was right next to, to loosen it up. Unbeknownst to me. I was up there checkin' all the wiring in them two rounds, make sure they didn't go. So, I went up there, checked it all out. They were just fine. Went and got lead line. Went back down the ladderway. So, this mess was in front of me. I said, "Gosh dang, that's not supposed to be there." So, I took it apart. That was to the bomb in the chute that I didn't know was there. So, I wired in these two rounds. Jaccard, who put the charge in there, was back at the box. So anyway, I get down. I tell the boss, I said, "Okay, they're wired in." He said, "Go back and help these other guys." There was 8 of us down there. Anyway, Jaccard shot. But he shot the two rounds up above that I had wired in. Not his bomb in the chute.

He thought he'd killed me. The boss about got 8 of us killed by doin' this. He was supposed to take care of all this stuff at the beginning of shift. You do not put two different guys out there. You shoot rounds, and then you go ahead with what you have to do the rest of the day. That's Federal Law. Period. Short and simple. (sigh) Anyway, we were back there workin' and the only thing that saved 8 people that day was my round did not go clear through. That much ground, about a foot of ground, saved 8 guys lives. That's the short and long of that.

Was the boss excused after that?

Yes, he was. He was blackballed. They sent the Federal Investigators in for this blasting incident. There was four of 'em, all with their secretaries. They took everybody on that shift, one at a time, and they took their statements. Everybody told them what the situation was, except for the boss. And the boss lied. So, they cleared all of us, and they blackballed him. He's not allowed to mine in the United States, anymore. Period. Yeah, that's what a blackball is by the Federals.

How did the Federal Government know about this incident?

You have to report that, yes. Oh, absolutely. Everybody was talkin' about it. I mean, there was a bunch of scared people. They were furious. Yeah, I was infuriated. That's where I got my nickname, "Wild Man." When I got mad I was anything but a very nice person. I was standing at the gate that night waiting for him with a two foot drill steel. And I chased him. I was about ready to kill him, because he lied. That's one thing you don't do in mining. You don't just get in there and just tell a pack of lies about somethin'. No, especially when it comes to guys lives, and all this stuff. It's one of them situations you just, I mean you just, pretty soon you're just red in the face and you're ready to kill somebody. It's very true. It's very true.

One foot of rock saved 8 guys lives, and I was there to witness every bit of it. I was up on top of the timber, we was puttin' in timber so we could drop the rock down, right? When that went off, you never seen any guys move so fast in all yer life, Man. It was whoosh... Gone! Oh, yeah, we were gone. I mean everybody was hittin' the decks, we were skiddin' under ground, under rock piles, just tryin' to get under somethin'. I mean, it was that close. Everybody was shakin'. Everybody was shakin'. Jaccard came back, he was white as any piece of paper I've ever seen. Thought he'd killed me. The first time I ever got hugged by a guy like that. My, God! He was cryin'. The tears were rollin' down, Man, I mean, he was so glad to see me. Jaccard, he's a Swede, anyway, I said, "Hey, what's the matter? You know you can't kill other Swedes." And I got him laughin'. He just dropped to his knees. He just dropped to his knees. It was just one of them real special times. Every guy there went and had a break, to say the least. I mean, okay, it's time to shut 'er down. That's it. We all went to sit down on the bench and talk. Oh yeah. We all went and talked.

The boss was there. He was talkin' to us, and Jimmy Junior told him, "I told you what the Federal Law says, now look what happened." And everybody says, "Yeah." Junior was right, and that was it. I mean, you're either right or your wrong; in a situation like that there's no in between. So we all sit there and "Well, we made it through that shift," and we went home. Then the main boss of the mine, he called the Federal boys in. They came in about four days later, a week later or somethin'. They each took a statement and then they blackballed Fred. But we were all... Man, I was so damned mad. Damn. I can't believe that. Because when you get mad, especially, I'm just going to say this the best way I know how, at stupidity, and they're laws that are in place to keep this stuff from happening. There's no reason for it. It's just plain stupid. Hair raising.

Oh, one more mining story for ya, on the funny side. Since we got all that serious stuff out of the way. Me and Brian Hayes were down on 3300 of the Ajax. And we were muckin' sump, with a shovel. It'd just been a miserable shift. I mean, the water's comin' down. You have a pump down there to keep it from gettin' too bad, right? You have to keep movin' the pump around to keep the water down. Well, anyway, you have a screen on the pump so it don't suck a lot of rocks and burn itself up.

So, me and Brian are down there. The water's just comin' down. We're standin' in about 8" of water. I said to Brian, "Hey, you think we should move that pump?" "Oh, we'll wait for a little while." Okay. So me and Brian are in there muckin' the skip full of rocks, with shovels. Pretty soon, you know, you feel the water runnin' in your boot. And you go, "Brian, water runnin' in yer boot yet?" "No." This bein' the skip was in the middle and I was on this side, he was on the other side. He wasn't quite as far down as I was, see? Because I started lower than he did. He looks at me, and he says, "Gottcha!" (laughter) Rotten son of a gun. Now Brian, he's very stoic. When he said 'Gottcha' it was just so funny. I just couldn't help but laugh. I mean, he was just so straight faced, nonchalant, I mean, I'm sittin' there goin', okay, fine. So, he did. Couldn't get mad at him. It was funny as all get out.

Were you working at the Ajax when it closed the final time?

Yeah, I was. Me and Brian were runnin' a little slusher drift, along with Rollison and Gary Horton. See, me and Brian worked one shift, and they worked the other shift. We were driving this little drift out there, and we had a different boss then. "Here comes Jack the Whipper." That's what we called him. He was from Homestake in Lead, South Dakota. He's an okay guy. Not real bright, but okay. I'm just kiddin', Jack, don't get all upset, okay? (laughter) Anyway, we called him 'Jack the Whipper,' he was really an okay guy. He really was. Anyway, "Well guys, we gotta find a million ounces of gold, or they're gonna shut this mine down."

So, we're just running along this little slusher drift, and we start hittin' ore. Yeah. Started off with, give ya an idea, the first round was 27 holes, loaded with dynamite, and we barely broke it. We got back in there, it was 16 holes, half loaded, because the hardness of the rock changed, it was softer, and we could go quicker. Didn't need as many holes, didn't need as much time to drill this. So, it started gettin' softer and softer and softer, pretty soon we're back there and we got a ounce of ore in the face. Well, here comes Jack the Whipper, says, "Oh guys, gotta back up and put windows in." We said, "Why is that?" He says, "Well, I was told to do this." I said, "That aint no sense, Bud." I says, "You want a million ounces of gold, we gotta keep goin' straight." He says, "Oh, no." So we back up and put these windows in there, just rounds, small rounds. Six foot deep, here and over here, and over here and over there. The gold aint there, folks. It's straight ahead, where we were goin'. Cause the rock was tellin' us that. So, they shut the Ajax down and we went, "Well, guess we'll never know if there's any gold back there, Guys." "Nope." That was Hecla. That was the end of it.

What did you do when your job ended and the mine didn't start up again?

Well, I always could find a job. I never had a problem with doin' that. So, I went to work for the Narrow Gauge Railroad, for Birmingham. Went out there and played on the track, doin' that. Then they started up this heap leaching deal. So, I worked at the mine then, too, but it wasn't much fun. I was road guardin' and worked on the pad crew, and done this and done that. Worked 6 months out of the year, then the other six months you were laid off. That was really hard on the family, I'll flat tell ya. Because the money was scarce, it was really hard. And, you know, I'm not really a Mr. Mom. Not too good. But we made it through that. I went to work doin' other things. I went down to Colorado Springs and worked down there. Then came back, worked on the crusher. Then I went to work for the City, that's where I've been ever since.

You know, I will flat tell you this job I've got now is the cleanest job I've ever had in my life. Well, you know, all mining jobs, they start out gold and they always turn to shit. Now, this job, it started out as shit, literally, the sewer plant is shit. But, it's turned to gold over the years. Put that on there.

Colorado Gold King Mine

Widow Makers & Hand Cars

"Underground is such a strange thing." ~Henry 'June' Hack

Well, I was born on a farm nine miles southeast of Parsons, Kansas. That was 1916. Around 1935, we were down in Springfield when the Dustbowl was so bad. There was nothing that would grow down there; the dust would just blow something awful. During the daytime it'd get so dark, you'd have to turn the lights on. You could see it comin' rollin', it'd come in. My dad had asthma and we had to leave down there.

I'm not sure just when it was he left Cañon City and came up to Victor. But he barbered there in Victor for oh, I don't know how long. I finished school at Victor High School, and worked off hours as a butcher.

My first time underground was over in Victor, at the Logan. I worked on the Logan for a while. And walked into a manway and got a bunch of rock on me. The Cresson was the last place I worked, and that was 1953.

I worked over on the Cresson quite awhile, run the hoist over there and I mined. Oh, did about everything on the Cresson a guy could do. They'd always find something I could do. I ran machines, the machine shop, the hoist. The last day I was there I pulled 555 tons of rock outta that thing; 65 skips. That was quite a place. I worked there until 1953, and then the grocery store down here on the corner came up for sale. We had that for 25 years.

I worked in the Portland mine a little bit. But I had quite a few different jobs there on the Cresson. I ran the compressors; they had five big compressors, and I worked on the skip. Take the hands up and down. Leased for awhile, worked for John Hanson, he was a lessor. I worked for him quite awhile. At the Cresson I worked for the Company most of the time.

I worked contract with Margaret's dad. He told me to marry her to get her off his hands. Then when we decided to be married, he said, "Don't take her too far away." She was the only kid. So, I moved her from across the street where our son lives, to over here. That was her home over there. So we moved over here.

At the Logan Mine, I was just a mucker. Fifteen cars or so, a night. Usually at night. They'll hold around 1700 pounds of ore, and I used to muck around 13, 14 a night. And tram them to the station. They were tram cars, you'd just push them. Later on in life they got these air machines to pull these cars around with, you know.

Over at the Vindicator they had these mules, but Roosevelt made them take the mules out of the mines every so often, give them a break. If there's anything that's pretty hard to do, it's put a mule in position where he could get in a skip and take him down underground. They're smart; they don't want to go down there. They took good care of them, fed them and electric lights all the time. They'd keep them down there quite a few months and Roosevelt comes along and says you can't do this, you won't. You've gotta bring 'em up on top every so often. So they did away with mules.

We just did it all by hand. We had power shovels for muckin', once in a while a company would do something like that for themselves. But they were expensive machines, and us guys would have to do the work by hand. Most of the things are done by blasting. Blasted into a stope and then the ore would come, and you'd run your track in under it and you could take it out. I could tell ya all about this stuff. But you drive a drift in and if you find ore, then you back up and make a manway on each side, put two or three chutes that are however long size. Then put the stope in, take the ore out and dump 'er in the pocket.

We had a water drill. The jack-leg was a later thing. The Leyner and the Stoper were big heavy things, but the jack-leg was easy to move around. Before that, these things, it takes about 16 holes, four foot deep to pull that down. It was a day's work. Kick that back onto plats, an iron sheet we'd put down to make it easier to shovel. That's how much you could gain every day, about a full sheet. The jack-leg didn't come along 'til I left the mines. We had the Leyner and the jack-hammer and the stoper. Call them the Widow Makers.

I was livin' in Victor at the time, and working on the Logan Mine. At fifteen minutes after midnight I picked up a bunch of steel and started out to the shaft with it, and the guy set the dynamite off. I was stupid enough, and young enough, that I didn't realize what it could do. I got about partway toward the shaft and he set the blast off, and it blew my light out. I thought I was standing in front of one of those chutes, but I wasn't. I was standing by the manway, and I got a whole bunch of rock down and got a cut on my head. I was spooky as the dickens for a long time after that. It could'a killed me, just as easy as not. I was just lucky. Friday the 13th is when it happened.

We used dynamite. You drill your holes and depending on the type of rock you've got, you put five or six sticks of dynamite and tamp them in there. And the long fuses on them, you know. You cut each one about this much shorter than that one, and you know how many you put in there, and you set 'em off. Usually it's a round of six to eight inch, eight foot fuse. Then when you get all that lit, you went down the manway and out, where it's safe. And you count them, to see if they all went off so you didn't have to worry about missed rounds. Usually about a dozen holes, you know. Course that's dependent on where they were in a drift or a stope. A stope you get as many in there as you can to break as much ore as you can. But the other one is to get a little farther each time, lookin' for ore.

We lost a friend, John. He had taken a tramcar and went back in this drift and he only got about ten feet in there and the gas hit him. Nobody would go in and get him. Hell, they could have held their breath. That's on ten hundred. Another guy got killed on the Cresson. His mistake was he went back in to see what it'd done, and the damn stuff was still breakin' loose. It caught him, and killed him. I remember them bringin' him up, and his leg hangin' down. That shakes you up a little bit, you know.

Did you ever worry and not want to go back down?

Oh no, it didn't particularly bother me that way. I could do this stuff. I didn't particularly like it, but I could do it. I worked in the machine shop, the blacksmith's shop.

What did you do at the blacksmith's shop?

Oh, mostly keep your steel ready. That's when you were using steel you had to keep sharp, before we got carbon inserts that would drill ten times as much as doin' it by hand. Dependent on the rock. Underground is such a strange thing. Over there on the Cresson, they got this damn dyke that runs across. It's a black dyke, kinda clayish lookin' stuff and it often had gold in it. But you'd have a streak of maybe basalt, and on this side would be a brecciate, and just maybe that thin of a thing between them was what you were after.

We mined all the way around that, but they were ahead of us with that Vug. It was straight east of the shaft, about 300 feet, somethin' like that. It was the funniest thing, I turn around a quarter of a turn and I can't get away from it. My directions underground, I really have to stop and think what I'm doing.

Where there a lot of jokes underground?

I was up this stope with John Hanson's outfit. My brother Charles and another soldier came to go down the mine. They'd never been down underground, so little tricks were pulled. I didn't say anything about that air tank. That air pressure valve, with 120 pounds of pressure, you pull that valve open and squeeze, it makes a helluva noise. I didn't think anything about it. Scared the hell out of them. We was up in this damn stope, up about 50 feet probably, and Charles says, "It was alright goin' up, is there any way to get down from here without the ladder?" No, you have to go down the ladders. I said, "Bring that sack." We had a powder sack hanging over the wires, and it'd shock ya every time ya took a hold of it, see. He said, "Wise guy."

What do you call the frame sitting over the shaft?

Well, they're 'Gallas Frames;' that's all I know. I think over there in Victor, the steel, did you see that? That used to be on the Cresson. That was originally on the Cycle, the Golden Cycle, and when they closed the Cycle down, they took it and put it on the Cresson. Most of them were made out of timber you know.

There was quite a pile of mines when I first come up here in 1932. There was around 2000 people in Victor, and about 2500 in Cripple Creek.

Tell me about high-grading at the mines?

Well, see, they got to where it wasn't like it was in the early days. Say somebody would hit a good drift or something, had some real good ore, they'd give you a piece of it. They weren't fussy about it. In the early days they did, but they were getting $3 a day, and I don't blame them. $4 was terrible. But it's funny, at that time of our life, $4 would buy as much as $20 today.

What was it like to be a hoistman?

Well, 'course it's a lonesome job. You sit up there on a hill. We had a room at the ol' Cresson. You never saw anybody all day long. All you'd hear is the bells, and the flashes of the light. Flashes to tell ya what to do, you know. You wouldn't see anybody. Send that skip down there 1800, 1900 feet, and you'd have to be within a spot of a couple inches. They usually had little marks on this diagram wheel as it comes around. Each one of those, usually they'd have a little mark on the drum or something to show you where the skip was at. There's a bell signal on every level of every mine.

Did you ever have any problems when running the hoist?

No, I don't think so. The guy that was ahead of me on hoist, he took her, he thought he had one more left to go, on the drum, and he pulled the skip up into the gallas frame, and the sheaves. And there was no bonnet on it. Anyway, your sheave cuts the skip off of it.

We were down on 1200 one night and we flashed for the skip and the guy gave us the signal, but they wouldn't come down and get us. The whole mine was down there. So after about eight o'clock, that old skip came rattling down the shaft, you know. They wouldn't pull us out, but they'd let us down. So they put us down to the 1700 of the Cresson, which connects with the Ajax. So we walked across that. I'd have gotten lost a hundred times walking over there. You never saw so many tunnels and drifts, as they got under this damn ground. The hoist was out on that side. It had everybody worried, but they got it all cleaned up, and straightened up. The man had been hoisting over there for a long time; I don't know what happened.

I worked there for Margaret's Uncle. One time he just got off the skip, I don't know what level it was, and walked away. Looked around and the skip was gone. She'd come loose. Why it came loose, or how it came loose, I don't know. But it went all the way down to the bottom. That was on the Cresson. That old skip laid out there on the ground, where they'd brought it back up there, it laid out there for a long time.

We made a little money there, but I don't know. Just a lot of it was luck, you know, to find the damn stuff underground. The way it is today, they take the whole mountain.

School Section Mine Ore House

Tour Guide as Timber Man

"I was working up at the Mollie Kathleen as a guide and another fella came up and said, "Would you like to work for us?" ~ W.E. "Ed" Grosh

My first experience with the El Paso Mine was 35 or so years ago, when Dale Weaver was running the place and I went for a tour underground. I was 40 years old. I was really impressed with it.

Quite a few years later I was working up at the Mollie Kathleen Gold Mine, guiding tours, and of course, I had worked in a lot of the mines around here, too. The New Hope Tunnel, and we called it the Maybe Never, but it was the Mary Nevin. Oh let's see, there were several of them I worked at. I did most of the carpentry work in the Gold Bond.

I was working up at the Mollie Kathleen as a guide one year when George owned it, and another fella came up and said, "Would you like to work for us?" I said, "Well, what's the deal?" He said, "We've got to re-timber the shaft." It had collapsed the upper 30 feet. "I was told you were a good mine carpenter." I said, "Well, I'm pretty good."

So I went to work for them, and I had three or four guys working for me. We started re-timbering the shaft. Well actually, we didn't get to put a piece of timber in, we were just cleaning out the hole, and they ran out of money. So that left us without work. But I worked there for oh, six or seven months doing that. That's about all that I've done at that mine.

But I did do a lot of exploring underground. There's another way of getting underground in that mine, besides going down the shaft. You go down the hill close to where the Mary Nevin is, and there's a tunnel. You could go in that tunnel and get over to the El Paso at the Jewelry Store Stope. That was a huge stope, and story has it, and I'm sure it's true, when they discovered that in 1894 they hired 200 miners, and 200 miners worked on that one stope for ten years taking all the gold out of it. There was silver in there, too, of course, because silver and gold usually run together. But one tram car, one 1 ton tram car came out and they figured that it produced $72, 250 worth of gold. That was when gold was $20.67 an ounce. That was a very, very rich stope and the El Paso was one of the ten outstanding producers of gold in the District here.

The Standard Tunnel was the name of that tunnel, and that was really neat. You go in that tunnel horizontally and just before you get to the dead end, there's a little tiny opening, about, oh, maybe three feet in diameter, going upwards. You have to get on your belly and crawl up through that and go up about ten feet. Then you can stand up in a huge, huge area that has been mined out, which was part of the Jewelry Store Stope. It was so fantastic. I've been in a lot of mines around here and I think that one is the best one. If you could get there again, I'd love to take ya. I don't think anyone's ever going to be in it again. But I used to take some tours, just not too many of them, three or four tours with the approval of the owners, and we had a whole bunch of lanterns there. So I gave everybody a lantern. That was all the light we had, you know, and it was so impressive to see all these people and their lanterns bobbing along. (laughter)

Was the El Paso tour run with lanterns?

No, no. They had electricity in the mine in that front area. When the shaft collapsed the power company cut the electricity off.

Mr. Weaver was running tours in there, and I was in the Volunteer Fire Department here in Cripple Creek. This is before I went to work there. And the fire started. I don't know how it started. There's a lot of rumors, I won't repeat the rumors because they're just rumors. But the fire started, and burned the mill. Then it was a few years after that the shaft collapsed. It was starting to pinch at the top. The top 60 feet of the shaft was just fill rock. They had built it up, and that's why it was squeezing in on the timbers to the point where you couldn't get the skip all the way up. So, somebody shaved the timbers off so the skip would go through. Well, then that made them weaker yet and then somebody went with a little crawler, you know, a little bulldozer, and tried to move the rock away. Made the mistake and caved it in.

One thing about that hoist there, it was a double-drum hoist. While one skip went down, the other one came up. They were counter-balanced like that. It was a huge hoist and a very good one, too. The operation of that hoist, the muck rock and the ore, was a little bit different than most mines. Most mines when you're underground and you mine ore, you dump it into an ore car, and then one way or another it goes back to the shaft, dumps into an ore skip and takes it up. Then they dump again on the surface and take it where it's going. But the El Paso Mine, they took the whole ore car up and they had room for two ore cars on each skip, so they were double the capacity of most mines. Not all the mines, but most the mines around here. And they would take the whole thing up. Way up to the top, they'd dump it down the chute and it would go right into the mill.

The mill was right next to the shaft, on the north side. The top was kind of covered in like a little room, because winter-time, you know, it gets awfully cold. The guys would have to climb this ladder on the outside to get in the, I guess you'd call it an ore house. It was very small, just covered in with corrugated steel. So they'd work up there.

Well, when the fire happened, everything collapsed except the little room on top of the gallas frame there. So, when I went there, before we could start repairing the shaft we had to take that down, because that tin was just flopping in the breeze. The ore house was on the east side. Everything's gone but the shaft. We had to take that tin off to start working on it. That was kind of scary, to get up there, you know. In the late spring, the wind was blowing and it was flapping off. Every once in a while things would fly off somewhere. (laughter)

We were there; the Fire Department was there, but all we could do was let it burn. It was too late when we were notified, to do anything about it. All the wood superstructure was gone; nothing was left but a pile. It was arson; the fire department believed it was arson. But who the arsonist was, nobody knows.

Like I said, it was a very big producer. Really neat underground for a tour. At one time there were two tours here. The Mollie Kathleen and the El Paso. The man who ran the El Paso Mine was a rather religious man. He wasn't open on Sundays, and he advertised that he wasn't open on Sundays because it was God's day, and that's why.

What about work at the Gold Bond?

That was for Al Goodwin. I did the carpenter work there. Built ladders and things like that for them. I never did do any actual mining, but I was underground a lot there.

At the Mary Nevin, I worked underground for a while. Then they were short of help one year, so I'd work a shift underground, and then when I came up I'd work a shift in their mill. It was nice getting two paychecks every week for a while there. But cold weather was coming in, and the water froze. We couldn't get water to the mill. So, that just about shut it down. They were still doing some mining underground, just very marginal. Not very much. Then by the time good weather came, as I recall, I think they just gave up mining. The people that owned it decided they didn't want to put any more money into it, because they hadn't hit anything worth while. They just shut it down.

Ben O'Haver had his own thing with a turquoise mine and what they called the New Hope Tunnel. I worked for him one fall laying track. This was right next to the turquoise mine; the New Hope Tunnel was the beginnings of a gold mine. Ben owned it; he never knew what was in it. So, he took samples and he was pretty sure he had enough to go ahead with it. He hired me to lay track in it, so the next spring they could come back and start mining. But that never happened either, and I don't know why. I laid quite a few hundred feet of track there. It was an existing tunnel, and it was in good shape, too. But there was a puddle in the floor, and Ben and I were walking in and Ben grabbed my arm. I said, "What's the matter?" He said, "Don't step in that puddle." I says, "Why?" "It goes down a hundred feet." (laughter) So, I built a little bridge over it. That was funny.

Did they have problems at the Mary Nevin with the shaft freezing?

Yes, they did. Sometimes in the winter-time you had to chip the ice out of the shaft so you could get the skip to go down. I did that a couple of times. I thought that was neat. We went down the ladder-way because it was fairly close to the surface. And we'd just chip it out and let the ice fall. There was a pretty good sized sump at the bottom there. Yeah, that was kind of interesting. There's a lot of stories about ice in these tunnels. There's one mine, I can't remember the name of it, but it froze over way back in 1920, and no one's been in there since. Ben and Jack, I can't remember Jack's last name now, but they decided they were going to blast it out. Well, you can't blast ice. Not very well. They tried a round, and it didn't work. So they took some mirrors, and tried to shine some sunshine on the thing to melt it. I don't think they ever got in there.

Why doesn't ice blast well?

Because ice gives, and it gives rather than breaks.

What do you think about Tommyknockers?

Well, you know I am a modern person. I know the old timers believed in the Tommyknockers, and Steven King had no idea what a Tommyknocker was, but they are like cousins of leprechauns, only they were underground. And the Tommyknockers kept playing jokes on the miners. He'd tie their shoe-strings together, eat their lunches. They would, when nobody was looking, you know, they would take a stick of dynamite, break it open, and rub a little bit on the headband of a hardhat. Then of course, when you put on your hardhat, about twenty minutes later, you've got a headache like you never had before. Nitrate headache.

Did you ever get one of those?

No, but I came close. I caught the guy. (laughter) It wasn't a Tommyknocker, by the way. We played a lot of pranks underground. Some of them were, I would say, dangerous. But not in our eyes, you know, they were just pranks. We wouldn't hurt anyone. Sometimes you'd drop a blasting cap with a fuse in it, about that long, you know, where nobody expected it. Bang, it'd go off and scare everybody.

There was a story at the Mollie Kathleen Mine. Years and years ago, like maybe right after WWII, when they took a tour they'd go down to the 7th level and take a sharp turn there right away at the shaft, and go back to the Queen Beth Stope. All the guys had a pocket full of blasting caps, and some fuse. And they would demonstrate how to put a blasting cap on a piece of fuse. Then they would light the fuse and drop it down the hole. It'd go down, maybe a couple hundred feet before it'd go off. It apparently was pretty impressive. This one fella forgot to let go of it. Took off his thumb and two fingers. And he was so dedicated to his job, he just put his hand in his pocket and finished the tour. (laughter) I don't think I'd have been that dedicated. That's the story that I heard; everyone said it was true.

What did you think of the ride down the skip the first time?

It was thrilling. It was thrilling, yeah. Certainly was. Course, I'd worked construction for a lot of years before I came here and some of the tall buildings, we had outside skips, or elevators. They called them elevators. It was very similar to that, except it was really closed in on all four sides, instead of just one side. Yeah, it was a thrill for me. A lot of people 'Ohh' and 'Ahh.' A lot of people say "Oh, dear!" A lot of people get scared going down. But once you're down there, it's not so bad. You walk around and the next problem's getting them back up. Some people... and you know George, he'd load the cage, and we only had one cage at that time. He'd push the people in and someone in the back would say, "I'm scared, I gotta get out, I gotta get out!" He'd send it down anyways. (laughter) Once in awhile one of those people didn't want to come back up. "Well, Lady, you've got to." I've talked quite a few of them into getting on. Holding my hand, or something like that.

Was there high-grading?

Yeah, high-grading happened a lot in the past. It still happens. Every mine I've ever worked for, if you came across some really good looking ore and you told the boss, "I'd like to have a sample," he'd say, "Go ahead and take it." But they know you're not taking it to sell. Real high-grading was people that snuck down at night or they'd have a deal going with the top hand. The top hand was the man who dumped. When the ore came up in the ore bucket up into the headframe, then it would automatically dump down into a little slide to the tram car. Well, some people had a deal with the top hand if they found any really, really good ore they put it on the bottom and then put the other ore on it. When it goes up and it dumps, well the bottom is on the top, then. The top man would see that and as he's pushing it over to the mill to be processed, he'd drop it off on the side, see. Then he'd come back at night, pick that stuff up, take it to an assayer, and some of them were not very honest. And so for a fifty-fifty split, he'd draw the gold out of it and you'd get half and he'd take half. I know that's the way it worked, because I was told that's the way it worked, by people that did it.

Then there's a story about an old timer who refused to leave his little shack. He and his wife refused to leave his little shack up on the hill there, where the mines were. They finally just forced them out of their shack and they found a little tunnel in this house down into a mine. It didn't go very far. He was high-grading ore out of there. That's how they lived. He'd take it, go to Denver every once in awhile with a bunch of high-grade ore. They'd get two, three, four hundred dollars. That's all they needed, you know. It was Lil and Sam by Midway. About that elevation, right in that area. That's a story I heard, you can't prove it by me. She used to stop in at The Grubstake where I was tending bar. Lil would come in every day. She'd buy a drink, and I'd give it to her. And I'd give her another one. She had two shots of bourbon everyday, and always gave me a dollar tip. Every single day.

Tell me about riding the skip?

Hoistmen all have their little quirks. One of them will start you down, they start you down slow and then all of a sudden ... at the Mollie there, it's an electric hoist. It's got a rheostat, this way is down, this way is up. And they bring it back like this and just let you go down slowly, and then just push it all the way back. Then go pretty quick, for just a second, (sharp intake of breath) one of the guys always does that. He doesn't even know he's doing that. Another fellow will drop you down pretty decent, but he doesn't slow it down 'til at the last minute. Another one will always drop it and it's too low, even after you ring the bell to stop it, he'll drop an inch too low and have to bring it back up again. Things like that. Sure, that happens a lot. You can pretty well tell after you've ridden with a few just who's running the hoist.

What would you say about underground mining, what would you tell people?

Well, it's a very hard job. 'Course there isn't many places you can work underground anymore because of the ecological impact on the atmosphere from processing the ore. The open pit mine here is about as safe as you can get. It's much safer than any other method ever has been. Not drinking cyanide or anything like that. It's all surface mining and you're going to learn to drill and blast and drive a truck, and you're going to be called a miner.

The last mine I remember that I was associated with that paid pretty good wages was in Nevada. I just spent my vacations there, because that's where my ancestors were from.

The Grosh Brothers discovered The Comstock Lode in Virginia City, Nevada. Yes, indeed. They had to go to California over the Sierra Nevadas to file a claim. They had a pretty big gold strike, but summer was the big thing. So one of them said, "Aw, let's just work another day." And he hit his foot with a pickaxe, got blood poisoning and died. Well, the other brother stayed with him until they buried him, and by that time it was too late in the season to go across the mountains. But he decided to go, anyways. And he died; he froze to death. But before he left, he hired a drunken bum from a bar, just to watch the place, live in the house, watch his claim 'til he got back. Well, this fella heard they were both dead now, he took the claims for his own and he sold them for $10,000. Six years later, when he found out they were taking $10,000 worth of silver a week outta that mine, he committed suicide. His name was Henry Comstock. That's a true story.

Hoosier Mine

Thanks a Lot, Bastard

"I was pretty up front about the fact that I didn't know much. They put me with this old timer. He was just an old drunk, but he was a pretty damn good miner." ~ Gary Horton

I was up here after a falling out and needed to work, and the Ajax was hirin' hands for underground samplin' and surveyin'. So, I hired on and started workin'. Back then they were drivin' the Cresson Lateral, up the 3100 of the Ajax, and puttin' in a new hoist. The hoist that's up there now was put in there in '75. So, I did some surveyin' with Vic, he's a local; grew up here. And worked with Chip, Shorty, and Steve.

We were workin' samplin' back in the Cresson, Vindicator, Theresa and Portland. We'd work off of the 2000 level. Off into the Roosevelt and start takin' different trips and what not. Just explorin' for the geologists. They wanted samples, so they needed somebody to pack the 75 pound samples in a backpack, back to the shaft. (laughter) That's what we were doin'. Just human mules.

But you know, geeze, we got to see some neat stuff. We got to see the old donkey stalls and what-not. Yeah, and some of the big stopes over in the Cresson. We'd be goin' back into the 1704. It was called the 1704 Stope. It's just huge. I'd read accounts of it, about it havin' a suspension bridge across it.

But it's true, you couldn't shine your light and see the other side. It had to go from the 2000 level of the Cresson all the way up to 300 level. It was just huge, just absolutely huge. You could see some pretty nice pillars in there where they left some of the gold. You could shine your light up there on a pillar and see it reflect. But problem was, the stuff was 30 foot up there and you couldn't get a ladder up to the pillar to cob on it. But it was really neat back in there.

What did you think the first time you went underground?

I had studies in geology and what-not, so to me it was just fascinating. But that's not to say I didn't see a lot of guys get down there on the Station or something and wouldn't get off the skip. Just ripped 'em up. I think, though, maybe some of the other guys did... after you've been doin' it for a while, you can tell who's gonna be able to cut it, and who's not. You can see 'em at the collar, too. You know, before you even hop on the skip, they're just bundles of nerves. And I can't say we probably didn't encourage that.

What did you think the first time you rode the skip?

Back then they were usin' the old hoist and it traveled, I think the Federal law came around in '76 or somethin' like that, that said a skip couldn't go any faster than 800 feet a minute. But that one, I don't remember for sure, it was like 1100 feet a minute. You were flying, almost like a free fall. It was a kick in the ass, you know, seein' the lights of the stations kinda flying by.

But comin' back up, when you're hoistin', the rope wraps around the drum, right. When it'd get to this one point about the 500 level, for some reason I don't remember exactly, the wrap wasn't right. It wasn't a smooth transition to go back the other way on the drum, and it would 'slap'. When it would slap, you would kind of be forced down and jerked back up really quick. I always thought that was kind of unnerving. I'm sure that everybody was kind of nervous about it, but you don't want to show fear. It's a macho thing, you know. That's just the way that is.

What did you think your first time in the dry?

(laughter) Um... you know, you're just kind of the new guy. I really honestly don't recall a lot of tension. You know, there's a couple of guys, one guy in particular I hate, but he's about the only guy I can think of. I still hate him. I mean the big thing, the one thing you learn goin' in to it, I think, is you gotta earn their respect. You gotta know that initially you're gonna get all the shit work. That's just the bottom line. They're gonna make it as miserable as they possibly can for ya. But if ya work, and you work hard, you'll probably make it. But you don't back down. You gotta earn their respect.

You worked your way up from sample crew?

Well, not actually. I bailed and just like took off for about a year. I worked up in Idaho in the Coeur D'Alene Kellogg. Worked for Bunker Hill. I worked briefly underground, but pretty much ended up at the Bunker Hill Smelter. I was hired on, pretty much, just as a scab to work for the Smelter up there, because the smelter hands were gonna go on strike. I didn't know they were goin' on Strike, so when I hired on, you know, you're goin' around the mill and everybody's talkin' about the impending strike.

It was a steel workers union and I realized that I wasn't gonna have my 90 days in. So, they hired me to be a scab. I worked all the way up until the strike then I split, and kinda wandered aimlessly for another year. Then I came back up here pretty much for good in '78. I hired on with the 'Nevins. The Mary Nevins. $3.75 an hour.

Straight pay?

Yeah, and beer was a buck, so I wasn't gettin' rich. I hired on as a miner, even though I really didn't have much experience whatsoever, aside from just muckin' the piss ditch and what-not. Just back breakin' stuff. I hired on at $3.75, and then we were just above the 350 level. I hired on in the summer so I didn't have to chip ice.

Did they have a problem with ice?

Yeah, back before they opened up 270, or got down to 270. They were expanding the shaft to a two compartment shaft, because you had to have a manway and it was a single compartment before that.

No manway?

I don't recall there being a manway. We were slabbin' off to put a manway in, put landings in. For every ladder you put in a landing for the ladder, and another hole, all the way down. For plumb bobs we'd use a nut tied to a string. If it looked pretty close it was good enough. Sinkin' shaft is difficult, even if you're just slabbin' off, because you're workin' against gravity. You know, you're hand muckin' to fill the drum. We had a sinkin' bucket, which is just a 55 gallon drum.

Some days it would ice up so bad, because the winters were a lot colder back then. Sometimes you'd have an upcast, where the air is actually comin' from the mine up. It's ambient down to about 2000 feet, I think, and it's about 52 degrees all the way up. Lower it starts gettin' hotter. But if the air was downcast, it would ice up. There was water always flowin' in the mines, you know. It would ice up the shaft so bad, the guides, that the skip wouldn't go down. You always ran your shaft before you went down. Yeah, when the rope went slack, you knew somebody'd have to go down. Get in the bucket with an axe and just start chippin' ice, so you could get to work.

They'd lower somebody down in the bucket?

Yeah. In a 55 gallon drum. You'd be spinnin' around and trying to chop ice. (laughter and spinning in his chair) Yeah, there wasn't a lot of guys raisin' their hand, but there weren't a lot of guys that were workin' there either.

When 350 opened up, I don't think we ever had any really problem with downcast. You know, it was a deal where once the 350 was open, the 350 was opened up to the Lonaconing and the Henry Adney, and the El Paso, and a few of the other holes, so you were gettin' air comin' outta there. And a lot of times it was just a matter of, you know, blockin' off certain air flow.

You hired on as a regular miner?

Yeah. I didn't know shit about it, really, you know. (laughter) Here's a funny story: Me and Johnson were workin' down in the shaft. Johnson would probably tell you this story, too, because it really upset him. In fact, it still upsets him, I guess. Anyway, you drill some holes and they were like two or three foot holes. You're settin' there with a jack-leg in that shaft. The shaft was only about 4x8' so it was pretty tight. Pretty tight for two guys, anyway. But you'd drill a few holes, and then you'd load 'em with powder, and then you'd tie them all together with det cord. Then you've got your fuse.

So Johnson, he was tyin' in the det cord to all the primers and we're gettin' ready t' shoot. I was loadin' up the sinkin' bucket with tools and I just happened to throw the axe at the sinkin' bucket. I missed and hit the det cord, and it pulled out some of the sticks of powder that we had in the holes. Johnson just goes, "What the %#@! are you doin'? Jesus Christ, you just about blew us up!" And I didn't know. Yeah, Johnson, you couldn't 'a stuck a needle up his ass after that one. (laughter)

When you went to mine in the Mary Nevin did people help you learn?

Actually, I was pretty up front about the fact that I didn't really know much. They put me with this old timer, Emery. Emery lived down in the trailer next to the headframe, next to the mill down there at the Nevins. He was just an old drunk, but he was a pretty damn good miner. Slight, though. And a serious drunk. I worked with him for probably a month and he never let me do anything, you know, other than push the tram car or what-not.

I said, "Jesus, Emery, when are you gonna let me do somethin'? Let me get on that jack-leg, I wanna run that jack-leg." He says, (imitating Emery as an old and high pitched voice) "Well, if ya wanted to run it, why don't ya just get on it?" So I said, "All right!"

So I'm full of piss and vinegar, right, and I take the jack-leg, you know, and I'm runnin' the cob back and forth just to watch the leg go up and go down, and I was kind of like this (moving up and down with the leg). He says, (imitating voice) "I'll go get ya a steel." So he goes and he gets this steel, and the steel is bent. And he did it on purpose, you know. The worst thing about bent steels is when you actually start drilling, yer machine is goin' like this, because it's kinda followin' the bend of the steel. I didn't know any better.

I didn't know how to set my leg, or anything like that. So, he said (imitating voice) "Well, go ahead." So I hit the throttle, you know, and the machine starts twirlin' around, and it starts. I'm tryin' to hold the machine and I give it a little bit of leg, and the leg's goin' up and I'm kind of ridin' up with the leg, and comin' back down. I'm up and down, and up and down, and all the time that bent steel is wobblin' the thing. I realize, I didn't know it at the time, but my leg was too upright, and if your leg's too upright then the machine has a tendency to bounce back and forth when it hammers.

So I was just dyin' tryin' to hold on to this thing. I mean I was just gettin' beat to shit doin' it. All of a sudden the steel shank, it breaks. Like an idiot, you know, I hold on to the machine. I go flying by the drill steel, and it goes flyin' by my head by like 3 or 4 inches. I go flyin' into the face with the jack-leg, and Emery kind of waddles over to turn off the air. Once he stops the machine, and I'm kind of knockin' the cobwebs outta my head, 'cause I mean I like damn near killed myself, Emery said (imitating voice), "Lesson number one, don't ever hold on to your machine." I was like thanks a lot, bastard. (laughter) You could'a helped me.

We drove on that cross-cut for about 50 feet, and we had this kid. He was a Freshman from the School of Mines and he was there as a summer student. He was real proud of himself for bein' there. He gives us our bearing for drivin' that cross-cut, and as he's walkin' out, after he gives us our bearing, he says, "Ya know, if you'd finished college you'd know how to do this." It's just pretty much lookin' at a compass. So I said, all right, whatever. About a week later, after we'd driven about 50 feet on that drift, Mr. Twinem, that's all we called him; he was a mine engineer up here for years. He looked at his compass and says, "You know, you guys are off by about 11 degrees." Turns out that the School of Mines kid had forgot to set for his declination. But we hit some nice tourmaline down there. Some nice pegmatite. It's a real course grain granite.

Did the mill freeze at the Mary Nevin?

Just about everything happened to the mill. Every winter. We worked for a few months and then got laid off. Johnson fired me for takin' too long a lunch break. Then I got hired back on about three or four months later after I had sufficiently atoned. I think we pretty much worked there 'til about 1980. Then they just couldn't, Brad couldn't compete with what was goin' on with TG up at the Ajax. Couldn't compete for wages. I think there were a few hangers-on, but maybe about the same time me and Brian and Gene, we all kinda tramped over to the Ajax.

I worked on the C-7 project. I worked on the Carlton Tunnel rehab, off the Shelf Road. That was when Hecla owned it. Let's see, C-7, the 3000 level with Gary Owen, the 3100 we worked back in the New Market. That was interesting. We were doin' some square set timberin' through a big stope we were gonna drive back off into the northeast, on the New Market. Me and Ed, we went back in there to do some square set timberin' and what not. At the same time, there was John and Gene, and Roy, and a couple other guys drivin', workin' a stope up above us. They were supposed to hole through, and we're supposed to get this square set done before they holed through.

We were usin' electric back then, for explosives, and we had this guy, 'Killer.' He was this German from Nazi Germany; he was actually in the Hitler Youth. He worked concentration camps in Poland. Killer had been down in Nicaragua; after the war he came and immigrated to South America, along with a lot of other Nazis. He got into a knife fight in some bar in Nicaragua, and the knife penetrated his larynx, so he had like only half a vocal cord, or something like that. So he talked like this (imitating a throaty, raspy voice). He had a strong German accent.

Anyway, Killer was hired on as a shifter, and he was our boss. Roy and John were pullin' chute, and they'd shot a round the night before on the shift, but they'd used millisec delays, and it throws the round. Well, what they did was they shot millisecs and you run a trunk line up the, they were up in the stope, so they ran a trunk line, which is your basic electrical cable up there. Then you tie it in to that trunk line, then you went down and used an electrical shock box. A shock box that would send the electrical impulse, you know, up to your caps and explode it.

Well, Geno's round, the trunk line got cut, so it didn't go off. We were fixin' on shootin' that round at the end of shift. So, Geno's up in the stope last I knew. John and Roy'd seen him, and Geno had tied back into the trunk line. We had like three trunk lines runnin'. John and Roy are pullin' chute and they get this big boulder in there. And what we'd do, we'd just take a little bit of powder and just kinda paste it on there with a blasting cap, and then you'd shoot it, and that would, you know, break up the boulder so you could get it outta the chute.

Well, they tied in to the wrong trunk line. Me and Ed and Geno are back temperin' up on that square set, and it's supposed to be a hole through round, where everything would'a blasted on to where we were workin'. Man, you heard John say, "Fire in the Hole!" and we thought we're safe. We're down the drift a ways from the chute. Ties in to the wrong trunk line, set off Geno's round up in the stope, man. That thing, it just went "slap! whap, whap, whap!" We were up on top, up in this stope, or about 20 feet up on the timbers. I mean it was assholes and elbows outta that thing, just runnin'. You know, after the smoke kinda clears and everybody comes in, John comes runnin' in and there's Geno, you know, Geno's standin' there with me and Ed. John thought for sure he'd blown Geno up, because he thought Geno was still up in that stope. That ended up costin' Killer his job. That and drivin' off the road. Yeah, he was the boss. Boy, he was a notorious drunk, too. He's one of those guys, you know, that would piss himself. He'd get so drunk, he'd piss himself.

How long did you work at the Ajax?

Up until about '84, '85, when they shut 'er down. That was, I think in '83, they had the first big layoff. Because they had put on a lot of people, and Hecla came in, or I think it was just before Hecla, TG decided to lay off a bunch of guys. Because they kinda lost their ass, I think, on the C-7 project, the decline in the Cresson. I think it turned out that they'd surveyed wrong, or something like that.

For the most part, I spent 8, 9 years underground. After they shut down up here, I went and worked a contract up in South Dakota in the Black Hills. It was a camp job, you know, we camped out and worked. It was good money. But yeah, I liked underground, it was great.

I got hurt a few times. Almost got killed once. Over in the Nevin I got slabbed on. We were workin' on the Ball Room, about the 350 level, when we were slabbin' off. I didn't bar down very well, I guess. I was muckin' on the pile, and just outta the corner of my eye I just saw this slab comin' down. So I started runnin' and I ran right into the muckin' machine, and the slab came down and pinned me up against the mucker. Rich was my helper, and I said, "Hey, Rich, I need a little bit of help for something." So it turned out he kinda held the slab back, I think it was about a 3 ton slab, held that back and I lifted up the bucket on the muckin' machine and swiveled it around. It was heavy. It wasn't that heavy, but I swung it around and got outta that. I definitely got knocked on my shoulder, screwed up my shoulder.

Were you more careful at barring down after that?

Yeah, I was definitely more aware. And over in C 7, I got slabbed over there when we were drillin' support split-bolts up in the back. Some slabs fell outta the back, the roof. Back on that drift we were drivin', it was a 12x20' drift, using two machine jumbo and the stuff that sits there on that observation point, over by the ADR Plant for the mine. You know that little scenic drive? Yeah, that machine we used underground, on that C-7 project. A Young's buggy, and that little loader kind of thing.

That whole deal was supposed to be a block cave type deal, where you drive a bunch of raises and a bunch of drifts, and then you shoot all the pillars out and the whole thing is supposed to collapse. Problem was that when they went to cave it, and they shot everything, nothin' fell. Then they decided it's really kind of too dangerous to go back in there again. At least that's my impression; I don't know what the real story was. But there wasn't any gold in there anyway, it was dead. You know they based that whole C-7 project on one intercept of like half-ounce gold. Yeah, just one drill hole hit some half ounce ore for maybe three or four feet or something like that, they thought it'd be consistent. But, it wasn't. It was dead.

Then I think in '82, or '83, I was workin' down on the 3300 level of the Ajax, and this was like right after the big layoff, so it was TG. TG decided to lay everybody off, or just keep a skeleton crew after they shut the C-7 down. Basically the way that it worked, I was workin' graveyard, had a guy come over to the house and he says, "Horton, you need to come in on swing-shift later." I said, "All right." He says, "You'll go to work." I hadn't heard nothin' then. Got there, ya know, and was wonderin' where the hell everybody is. And they pretty much said, everybody was like dressed, and they said go in there in the lunchroom and if your name's on the list, go to work. The rest of you guys go home.

I got lucky; I made the list. I think there was, gawd, I think there was only, after they had 40 guys underground, I think there was only like 15 of us left. Anyway, I ended up workin' down on the 3300 level, and I was runnin' a slusher drift, fillin' the skips. Why I had to go up to the 3100 one day to turn off pumps and what-not, I don't recall exactly what we were doin'. Well, pumpin' the sumps out so you could get the skips down far enough to load. So, I went up there and all a sudden, as I was goin' from 3300 level to 3100, the skip just stopped.

I was by myself. The other two guys were back up in the heading. We were drivin' on the South Bobtail then. I was going, "Huh, I wonder why the skip stopped?" All the sudden, boy, next thing I know is my face is down in the bottom of the skip and the skip has taken off at like 30 or 40 miles an hour. So it went up, and the rope went slack and after it kinda stopped, it started goin' back down. But it started goin' back down at such a rate that I was up in the bonnet. So, I was flyin' at the top of the skip and then all of a sudden it bottomed out, and it started comin' back up, and I was still goin' back down, and I just slammed, I slammed the bottom of the skip. It did that, probably three or four times. So I just got slapped silly. All I knew is that, there's only three or four inches of clearance between the timbers and the skip itself, in any direction. So all I knew is man, I better curl up into a little ball and try to stay in the middle of this shaft, because I definitely could'a got easily sliced.

Once it stopped, the first thing I had to do was check to make sure I was all right. Well, my leg was broken and my foot, my ankle, it was broken 90 degrees this way, and 90 degrees that way. And I was wonderin', where the hell is my foot? Because I can't find my foot. It was busted up underneath me. So, you know, I reached over and I grabbed the bell, and I rang the stop bell. Duncan was hoistin' at the time. I stopped it. I started to think, well geeze, what do I do now? I can't ring a 9 bell, 'cause 9 bell is like a general disaster and shuts the whole mine down, and there was no way to get a hold of those guys up in the headin'.

It just stopped. But after everything kinda calmed down, Duncan raised the hoist up to the 3200, which was an old skip pocket that was kinda intermediary. But there was no way I could get to the phone, without havin' to get outta the skip and drag my ass over to the phone. That's another problem with the bell signals, is that you know, if you do a general alarm, a 3-5-3, or a 9 bell, those are pretty much general alarms and shut everything down, and the hoistman can't move the hoist unless there is a verbal command to do so. I figured there's no #@%& way I'm goin' back down to the 3300. There's no way I can call him on the phone. So, I just belled myself to the surface. That was a long ride, because I didn't know why the skip had done what it did. I just knew that, you know, better off just takin' yer chances and gettin' up to the surface.

The hoistman was moving the hoist without any signals?

Yeah. It's not a good thing.

What happened that made the skip do that?

Well, there's these things called guide-shoes, which the guides are kind of like angle irons, they run right on the 4x4" that guides the skip up the shaft. Well, these shoes are kind of these things to keep the angle iron from actually diggin' into the guide, or somethin' like that. So, it's kind of this tapered little deal, and there's four of them, on each side. Well, the top guide shoe had come undone and it fell down and jammed into the bottom guide-shoe. So that stopped it. Well, when it stopped it, Duncan kept hoistin'. He could'a been able to notice the Amp meter, because the Amp meter would'a pegged out. It measures your amperage, so if your amperage starts to go up, then your hoist is actually under some kind of strain. I don't know, I can't speak for him, but he kept pullin'. When it finally dislodged, there was so much tension in the rope that it was just a rubber band. I just yo-yo'd up and down, three or four times, until it settled.

But what was weird is that there are these things called dogs, which are a camming device, you know. When the rope would go slack, the dogs are supposed to clamp down on the guides, to stop the skip. Well, that never happened. So, they were faulty. The guide-shoes were faulty, and I just happened to be the recipient. So yeah, I got to the surface, you know, and kind of breathed a sigh of relief. But I was just kind of sittin' there in the bottom of the skip goin' "Help." But I didn't cry. (laughter) I was whimperin' a little bit.

Were you mad at the hoistman?

Well, yeah. You can probably second guess him. But I don't know. All I know is that I busted myself up pretty good. That was bad enough. So Duncan goes and he calls the Victor EMTs. At the same time, the Cripple Creek EMTs and the Victor EMTs got a call about a kid who had tripped in the parkin' lot of the Mollie Kathleen, and he was messed up. So Cripple Creek says, we're gonna respond to it. So Victor, I don't know what the deal was with Victor, but all of a sudden, after about 20 minutes, Victor decides they're gonna respond.

They'd gotten the two calls, and Victor and Cripple Creek thought that they were goin' to their respective places, but they were both goin' to the Mollie Kathleen. So after about 25 minutes I start hearin' the siren, and I'm goin, "Shew, all right. Here we go." So I hear the siren, and it's gettin' louder, you know, and it's goin' up the mountain, Battle Mountain, and you hear it really loud, and all of a sudden you hear it fade. And I'm goin', oh, no, what's goin' on? They got like half way up the Range View Road and the dispatch says, "Did anybody respond to the accident at the Ajax?" And all the sudden you hear the siren gettin' louder and louder. But it was like an hour and a half of just settin' there in the skip. Then they wanted to immobilize my leg and what-not, and I just said, "Get me outta this fricken thing." I was out for 4, 5 months.

Were there other accidents?

Yeah, Joe Cook, he got real messed up. He was workin' with Brian, that same slusher trench I was workin' on 33. We had these rocker bottom ore carts, and if you weren't real careful, if you overloaded 'em or somethin' like that and if you weren't real careful when you pushed 'em over, the bucket would just fly off. And that's what happened. Except Joe went with it. He just tore his face up. The rock down there, the granite on the 3300 is just like razorblades, it'll cut ya really bad. Yeah, Joe went over with the rocker bottom and just got tore up in the slusher trench. I saw him, you know, like a week after it happened, he was over at Donkey Derby Days. He'd busted his jaw, he was drinkin' his beer through a straw. And just tore up, his hands are all bandaged up, you know, his face was just a bunch of zippers. I said, "Jesus, Joe, I hate to see that happen to you." A 'better you than me' kind of thing. A month later, I was in the hospital.

What do you think of Tommyknockers?

That's a myth. The only thing that's weird underground is if ya turn off your light about between midnight and two, you can hear the earth creakin' if you're in a real quiet place. It's just gravitational effects, I think. Rock is very elastic, actually.

Commander Drum had me up in the scrolls the Mary Nevin. You know as the skip goes up it's got little rollers on it. It'd run up into what's called the scrolls; those rollers would run up into the scrolls and cause the bucket to dump. Then they'd dump into an ore cart up there at the surface.

Hey, I gotta tell ya about the Nevin, though. It's just a miracle none of us died down there because really none of us knew what the hell we were doin'. Nobody really did. Emery, he was the big guy. He was the most senior of all of us. Emery taught me a lot, you know, but pretty much I just learned it all by gettin' on the machines, and that's the kind of philosophy I've took ever since then, too. "Just let me on that pig." If you wanna learn somethin', you just go and you get on it, and get off when they tell you to. (laughter)

There was one time where me and Emery, he was always trashed before we went underground. I never really got trashed before I went underground, but I admit, if we were workin' swing-shift or somthin' like that, I'd have a couple 'a beers or something. But, geeze, Emery would be trashed, and I'd end up doin' a lot of work for him. But that was part of the deal. There's definitely like, policemen have this camaraderie that you just don't break, that bond, that trust, because you really do count on each other to save your ass. And it's true. Because no rock ever said, "Ouch." Lawrence was our hoistman when I was workin' with Emery, and a couple 'a times we said, "Oh, Emery, we better get outta here before one of us gets hurt." Just one of those shifts, you know, where everything was bad about half-way in to it. A couple of times we just knocked it off and went over to Zeke's (Bar). That was the old days.

Johnson got buried at the Nevins. He was kickin' muck down in a chute. I was gone that weekend. It was funny, because it ended well. We had ropes you held on to when you were kickin' muck down in the chute, because if you didn't have enough angle, gravity can't work. That's why miners always liked to mine up, use gravity to let the rock fall. And the structure of the veins in the Nevin were a little bit flatter than 45 degrees. So, it had to be a little steeper than 45 degrees for it to work, for the rock to fall. Well, it'd hang up, you know. Then you'd have to go in there, you'd kick the stuff down. But you usually held on to a rope, you tied yourself off.

I don't now how he did it, but he somehow got below the muck he was kickin' down. The muck came down, and buried him. So I guess they were pullin' chute and it's really kind of funny, because Bill, I think he ended up goin' to Cañon City prison, but he was in on the 410 level. The chute went from 350 down to 410, and I guess he was down there and he lit up a smoke. And all the smoke was goin' up, and Johnson only had like a little hole, and he heard this muffled "Put the cigarette out!"

I got back in to town and Marsha comes runnin' up to me, she grabs me, you know, "Oh my God, I'm so glad that you're okay!" "What are you talkin' about?" She says, "Oh, geeze, didn't you hear? I thought you were over at the Nevin and you were buried. Heard there were seven miners buried underground." And there were only 4 of us that worked over there. It was funny, just what kind of rumors were goin' around town. "17 miners buried, 4 known dead," you know. It was funny.

There was one time where Ron Myers, we had t' put a refuge chamber in the mine so if there was like a mine fire, or somethin' like that, or a cave in, you could get back into this refuge chamber and there'd be some food and a little cot, and what-not back there. To where you'd just hang out and wait for help. And it's like none of us would even consider it, really. You'd find a way to get your ass out, if you were still alive. It would be futile, it was only a chamber. It was only a hole about four foot wide and eight foot deep. Enough for one guy, maybe two if ya wanted to get cozy. But Ron, one night workin' graveyards or somethin', he gathered up a whole bunch of cigarette butts, we all smoked back then, picked up all these cigarette butts and threw them back there in the refuge chamber. He brought a couple of, you know, Playboys and had those layin' around in there. And Jinx came back, and you know all of a sudden word was out that the boss thinks we're all sleepin' on the job down there. I think they took the cot out after that and cleaned it up. (laughter) But there was just no way that anybody... you'd figure out a way to get out.

Did you do assay?

Yeah. Beginning '87, I started assaying. Fire assay. Ended up runnin' the lab for oh, 12 years. For TG, NERCO, and all of 'em. I was Chief Assayer for that whole exploration project that started this thing up. Nobody knows what assaying is.

What is it?

It's just figurin' out how much gold there is in a rock. Anyway, I must'a done easily over 100,000 assays, just in my tenure. I learned from Carol Roberts. Yeah, that was a weird deal. She broke me assaying, and I actually, everybody knew they were gettin' laid off, and I just went up to Carol, and I got down on my knees, and I grabbed her around the ankles, this is previous to NERCO, now it's still TG. It was over at the Carlton Mill. Me and Brian and Gene, and a few other guys, we'd been workin' the summer crushin' rock. Well, samplin' the rock crushers that Southway was runnin'. All the dumps for the initial, the first heap. Winter was settin' in, and I just grabbed Carol around the ankles and said "Please, teach me to assay. Teach me." I was goin' to night school when they thought there was this hope for me. So she kept me on. I worked for her for a couple of years, while I was goin' to school. Then NERCO came in and they fired her and put me in her place. I felt terrible about that for years.

What do you do now?

I'm back in mining. Now I'm workin' Environmental Department. I went back to school and finished up my degree. Got a degree in geology. I started it in '74, and finished this year (2004).

I love my job, it's great. The last year I spent really studyin' a lot of geophysics, so you're studyin' earth movement and seismograph. So, I'm in charge of the seismograph project we've got goin' here. And goin' to work my way into fiddlin' around with a little geochemistry. Just monitoring and doin' some test work on waters around here. Make sure the mine's not polluting any of the water. Staying in compliance. That's another full circle thing, because I spent 25 years just tearin' it up, or aiding and abetting at least, and now my goal now is to minimize the amount of damage outside the perimeter of the mine. I can't stop it, money talks. And it's the only ball game in town, really. It's good money. I'm just glad to be back, too.

Gold Band Mine

Tramp Miner to Mine Owner

"I was probably seven years old. We were rounding up cattle, so I got the flashlight and climbed off the horse and went in there and wandered around."
~ Rich McNeely

I was born October 25, 1942, in Reno, Nevada. I have a Bachelors of Science degree in Mining Engineering from Mackay School of Mine Engineers of Nevada. Named after John Mackay, who was one of the big four up in Virginia City, which was big in mining in the 1860s. He started the mining school in 1902, I think it was.

What took you to Cripple Creek?

I was offered a job by Silver State Mining Company to manage the operation there, the Ironclad claim, they called it. They had a vat leach operation. So, I went back there to manage that. That was in 1983. I was there until 1985, after NERCO took over the property from Silver State. Then I left NERCO, and came back out to Nevada and started up another operation for Silver State, for the Reids.

Your family has been in mining?

Yeah, my great-great granddad was actually an attorney. He lived in Silver Cliff in the 1870s, and he was pushing for Silver Cliff to be the capitol of Colorado. They ended up losing by one vote, so Denver became the capitol. So, if you ever go down to Silver Cliff, they have a museum there and all kinds of articles about my great-great granddad. I think it his name was James Theodrick McNeely. Something like that.

Anyway, so then when Silver Cliff started dieing down, Cripple Creek started coming up. So he moved to Cripple Creek, and he was an attorney in Cripple Creek. And he had, I think it was 10 boys and two girls. Most of the boys went in to the mining side of the business and they traveled all over the world. I know two of them went down to South America and were down there for quite some time. Three of them were killed in the mines in Victor. One was killed at the Strong, another at the Independence, and if you can get in to the Elk's Club in Victor, they have a plaque up there with several McNeely's on the plaque. If you go in, right in the front where they have all those names, kind of like the founders of the Elk's, there's I think two or three listed there.

One of the sons moved out west and that became my granddad. He kind of tramp mined all over, and ended up in Washington. That's where my dad was born. He came back down to Reno, prospected all over Nevada, and ended up running a barbershop here in town. My dad had had enough of the mining game, you know, because he saw what his dad had done. So he completely stayed away from it. Here I am back in it again.

I was raised on a ranch most of my life. You can only have one person running a ranch, and I was kind of the bottom of the clan. I liked the outdoors and I figured I better go to school and get myself an education, so I came up here and went to school. I had been out with my dad or my uncle running cows, seemed like every time we'd come across an old drift or something I had to go see what it looked like, so I figured, well, I must have a little bit of mining in my blood. So that's what I did. Joined up at the School of Mines and ended up getting my education. It's been pretty good to me; I can't complain about it.

Tell me about the first time you went underground?

I was probably about seven years old, eight years old, something like that. We were up in the Stillwater Mountains, which is east of Reno about 95 miles, maybe 100 miles. We were up rounding up cattle, bringing them down because of winter and we had a few minutes and were just sitting there, and I happened to glance over and see this tunnel. I always carried a flashlight because we might be out late or something like that, so I got the flashlight and climbed off the horse and went in there and wandered around. 'Course I didn't know what I was supposed to be looking for, but I thought it was kind of neat. It was cold outside and it was warm in the tunnel and I thought, well, this is pretty nice. I think that was about the first time. My family growing up in the mining side of things, my daughter, I think the first time she was underground she was about five, my son was seven. They both enjoyed going underground. So far, none of them have followed my footsteps, although my daughter thinks she'd like to.

It's not easy to be a female mine engineer?

Oh, it's getting a lot easier. The barriers have been broken down a little bit. Out at the mine we have at Aurora, Nevada, which we call the Esmeralda Project, we have a woman who drives trucks and runs loaders underground for us. The barriers are broken; you don't see the superstition that you used to see where if a woman went underground the miners had the superstition that there's going to be somebody killed that day. You'd have a lot of your, I think it was the Irish, if they saw a woman underground, they'd tramp right then and there. They'd leave. But times have changed. Really quite a bit.

We have students up at the school of mines now, women in the engineering field. It's getting a lot better and it's a good paying job. You come out of school and make almost $50,000 to start, that's not bad. Mining is actually the highest paid profession there is, at least in the State of Nevada. It's the second largest industry in the state behind gaming. Gambling, then mining, then ranching. The top three.

Before I got married I tramp mined for probably five years, something like that. Then I went back to school and got my degree, and I had my own contracting company for another five years. Then started working for big companies, and small companies, and finally went out on my own 15 years ago.

The first place after I got my degree, I actually worked at Phelps Dodge down outside of Clifton, Arizona. The name of the mine's Morenci. It's a large open pit copper mine; I worked there for a year or so. Almost two years. Then I came back to Nevada and ran a small underground operation south of Battle Mountain. Ran that for about a year and a half. Then I went out on my own for five years or so, and contracted primarily. I had a little contracting company. I worked in all of the Western U.S. Arizona, Colorado, Idaho, Nevada, California, you know, wherever companies wanted work done underground. Then Jimmy Carter became President and interest rates went up to 22%, so I sold everything and went to work for Freeport, north of Elko, Nevada. I worked for them for 3 years before I went to Silver State in Colorado.

When you did contract work, what did you do?

Basically I did about 3 different things. A lot of times you'd have the exploration people would want to get underground in some of these old districts. So I'd go in and open up all the workings where they'd cave, and stuff like that. Open them up so they could get in and sample it.

I drove some drifts for some people that wanted to drill underground. They had a target they'd defined on the surface, so I'd drive a drift or a decline or whatever they wanted, down to where they wanted the drill station to be. I'd have it all set up; all they had to do was drive the rig down there and start drilling.

Then I actually mined a contract where people wanted "X" number of tons of limestone, talc, lead, zinc or gold per day, and I'd go in and do the mining based on that. So they didn't have to buy the equipment and everything else. I already had it, so it'd save them on the capital side. Cost them a little bit more on the operating side because I had to make a profit, too, but they felt that was better than spending five or six million dollars on the capital equipment right up front. So they could do that as mining progressed. I had a crew, probably up to 25 people for the most part. Just depended on the job, the size of the crew. Sometimes I'd have two or three things going, I might be up to about 50 people, but normally it was in the 25 range.

Was it dangerous when you had to go into the old workings?

Not really. Underground you've got to be cautious, anyway. Kind of like defensive driving. You've always got to be looking for things that could hurt you. Don't be laying your hand on the shaft timbers when the skip's coming down and could cut your hand off. So yeah, the danger is trying to do something you shouldn't be doing.

I'd been doing it for a long time. I knew exactly what to look for. I think one time, in Bodie, California, which is a historic ghost town now, it's actually a State Monument, we were underground and we were trying to drive through this area where it'd caved. We got about half-way through, maybe two-thirds through the cave, and we started hearing stuff come down from up above.

So, I turned around, there were two guys with me at that time, I pushed them out of the way, and got buried up to my waist. Thank goodness it was all fine; I was just bruised, nothing broken. I had to dig myself out and start all over again. So, that's the closest I've ever come to something happening like that. You know, it's part of the trade. Everything was going fine, but you could hear it. You could hear boulders coming down, and you just beat feet and get out of the way.

Did that make you wonder the next time when you were working?

Nah, I went right back in. Probably less than an hour later we were back in there trying to get through again. Get the job done.

One thing about that job, once we got through it, I found a newspaper that was dated 1905. It was a San Francisco newspaper, printed the day after the big earthquake and fire. It had pictures and stuff like that. I gave it to the museum. It was underground, in perfect condition, like it'd been printed the day before. Because it hadn't had the sunlight or anything like that. Half-way dry, so it didn't get the moisture or anything, you know. I was pretty surprised. Normally when you go underground and open up things like that, you'll find candle holders or bottles or stuff like that. But not a newspaper. And one of such significance; that was a devastating earthquake and fire.

After I left Cripple Creek, my family stayed there for about another year while I was out here starting up another project in Nevada. Then they moved back. Our two youngest were born in Colorado Springs, while we lived in Cripple Creek. So, it kind of has a special place for us, the family.

I think we all really enjoyed living there. At that time there was no gambling, so it was quite a bit different from today, from what I hear.

I was back there three, four weeks ago and visited the mine, which is quite a bit larger than it was last time I was there. But that's okay, it's mining. You've got to go where the gold is. There's obviously a lot in Cripple Creek. Up until I was there, there was like 21 million ounces pulled out of the District; now it's probably closer to 24 million, 25, something like that. It's a big District, when you look at that kind of district, which was primarily underground until the 1980s, when they started doing open pits.

I have an Aunt who was born in Victor. She was actually the first licensed woman assayer in the State of Colorado. That was back in the late '30s. She did assays there, and her husband was a mine engineer out of the Colorado School of Mines. He worked underground at the Portland until they closed in '62, I believe it was.

Maxine and Earl Philips. They lived in Victor. She was a Brown, which she got her name because of the two girls out of the 12 or whatever it was, she was the daughter of one of the daughters, so she had the name of Brown. So, there's actually a house in Victor that they used to refer to as 'The Brown Palace.' Which sits, as you first come through that narrow little passage and turn right, there's a real nice looking house; a pretty good sized house, at least it was back then. I think it's still there. Anyway, that's where she grew up. Her dad was a tramp miner. They were rich sometimes, they were poor sometimes.

Did she have a lot of neat samples?

Yeah she did. As a matter of fact, I've got a satchel I haven't even taken out yet. All the specimens are wrapped, and they're all specimens of Cripple Creek. So it has gold, it has antimony, it has silver, lead, zinc, copper. It's an old satchel, probably 18" long, probably 8" high, 10" wide. So yeah, I've got all the different types. She wrote a lot of poems and most of them are about Victor and Cripple Creek. She actually published a book on it; it's pretty neat. Her dad ended up going up to the gold district in Wyoming, and loosing about everything he had. The gold was tied up in arsenic, and at that time they just didn't know how to get it out. Now days you can separate the two, but not too many people want you mining arsenic, either.

Where there ever accidents of people that were working for you?

You know, there's always accidents. But, knock on wood, I've never had a fatality in any place that I've run. I've never really had a serious accident to the point that people were maimed, or crippled, or anything like that. Knock on wood.

In my eyes, and my partner's eyes, safety is the number one thing. We will not do something that's going to get somebody killed. It's not worth it. Now you look in other parts of the world and there's places where a human life doesn't mean much. Well, I don't know if I could even work there because of their philosophy. We're very protective of our employees. We feel that they're putting their life on the line every time they go underground, and we want to make sure it's safe. Come out and see the light of day again.

But yeah, I've had... probably one of the most serious ones was a guy was running an underground loader and he came up on the surface. He was working night shift and he wasn't paying attention. He was getting ready to dump the bucket and ran one wheel up on a pile. When he raised the bucket, the center of gravity just laid it over. It went completely upside down. It had a roll cage on it, so he wasn't hurt in that respect.

But the roll cage collapsed a little bit and the way it had rolled he couldn't get out. It had broken a hydraulic line, and he was getting hot hydraulic oil in his eyes. He lost his sight for about a week, I guess, before it came back. They kept his eyes covered, you know, and kept putting fluids in it and stuff like that. After a week they took the stuff off and he was fine. But he has some burns down his face from the hot hydraulic oil. I think that's about the worst one I've ever seen. I've seen guys twist an ankle or stuff like that.

When you mined, what kind of drill did you use?

Well, surface mining, of course, we were using a down the hole hammer blasting rig. Underground we used Jumbos. It can drill basically from vertical up to vertical down, and any range in between. Like we're driving, out at Esmeralda, we're driving inclines that run about 15%, and so the drill is actually drilling a little sub-horizontal, to keep the decline going at the right percentage. Periodically in stopes or stuff that we're doing, we use jack-legs. We use slushers, we use front-end loads, we use trucks that haul underground.

Just about all the Jumbos are electric, over hydraulic. So you have a big extension cord that you plug into these things. That's 460, 3-phase. So you've got to be careful of shocks and stuff like that, make sure it's grounded. Just about all Jumbos are that way. Of course, loaders and trucks are all diesel so you've got to have a lot of ventilation to keep the particulate emissions from the diesel down. You've got all the MSHA standards you have to meet: noise, pollution, so on and so forth.

When you did contract, did you do it the same way?

Yeah, basically. For the most part I used jack-legs. The Jumbos really hadn't come in to their own yet. A lot of them were diesel, over hydraulic, and they just weren't that efficient. A lot of them at that time were just air, over hydraulic. So, times have changed. They're a lot better drill than they were 30 years ago.

We had a front end loader. Depending on what job I had, we either had rail haulage, or I had front-end loaders and small trucks. So it varied from job to job. If it was really level, the tunnel going in was level, I generally used track and a little over-shot mucker. It was air powered. If I was running a decline or something like that, I'd go ahead and run rubber-tired equipment, front-end loaders.

Do you believe in Tommyknockers?

Well, about half and half. If you have somebody telling you to get out of the mine, you'd better listen to it. Which Tommyknockers, that's what they do. And you know, I've been in underground mines where things just didn't feel right. You feel like something's wrong. That's when I leave. I listen to myself. Now, whether that's a Tommyknocker or not, I don't know.

What would you tell people about mining?

I think mining is a real good field, if people like the out of doors. What I've found basically is people either love underground mining, or they absolutely hate it. During the 1980s and early '90s if you tried to talk underground to an engineer coming out of school, they didn't want anything to do with it, because there are all these large low-grade open pit gold mines operating.

Now you see the attitude changing. We're actually trying to hire a kid that's going to graduate from Mackay in May, and it seems he really loves the underground. Like I say, I haven't found too many people that walk the fence. They either love it, or they hate it. I've seen people quit when they found out they were going to be taken from an open pit mine to go underground. They'd just say, "Forget it." So, I don't know.

Is it something that gets in the blood of people?

I think so. Underground is, to me, a lot more exciting because of the fact that every time you drill a round and shoot it, you know, there's something different. Maybe it's better ground, maybe it's worse ground, maybe you hit a nice vein, whatever. With the open pit, that's an earth moving proposition then. I think you loose a lot of that personal contact that you have with underground miners. I go out and talk to just about everybody on the job and you don't see that too often, especially in a large open pit.

You know, if you take an interest in your employees and treat them right, they're going to treat you right. Every once in a while you get burned, but that's okay. I'd rather be on the side where I treated him right and if he turns around and stabs me in the back, he'll never work for me again. Whereas 99% treat you right, if you treat them right. You've always got the one. One bad apple. It's not just in mining, it's every business that there is. That's life. You can't change your own philosophy for one person.

The first mine that I actually owned was Cobre Mining Company, down in Silver City, New Mexico. We actually had three mines. One open pit, two underground mines. We bought that out of bankruptcy in December of '92, and started production in the second quarter of '93. We ran it up until February of '98, and there's a story in there.

What we did was basically we took a mine that had been shut down for eleven years. There were two mills on site, and we went through them and we revamped all the underground workings and started this thing up. We basically took it from cost wise, the previous owner costs were well above a dollar a pound for copper. In the four and a half years we ran it, we took it from that dollar a pound down to .36 cents per pound. To concentrate you add about a sixteen-cent charge on top of that.

So, we were making copper in the low $.50s, the high $.40s, which was about the lowest cost producer in the U.S. at that time. Especially in the Southwest. We took the production and we tripled it. We increased reserves about five times. Obviously reduced costs. Then we took the company public in April of '97, and we'd been public about six months and Phelps Dodge launched a hostile takeover on us. That was successful; they got the company in February of '98. They ran it about a year, then shut everything down because they said they couldn't make any money there.

They had fired 19 of the top 21 people, then they couldn't run the mine. So, you know, you hate to see that because we had about 315 employees, and were making good money. The company was making good money. What we did was we would always take about 5% of the net and give it back to the employees every month, kind of like a bonus for making money. So they were making real good money down there. I guess, we were right up against Phelps Dodge property, and they didn't like us showing them up. So they took us over. The mine was called the Continental. And we had, the two underground mines were the Number 3 and the Number 2. The Number 1 underground mine actually became the open pit.

Which was the Continental?

They were all three, it was the Continental Project, is what they called it. It had been owned by United States Smelting, Refining & Mining Company, prior to Sharon Steel. And when Sharon Steel went bankrupt, we ended up buying it.

What are your plans now?

Well, we just opened up the Esmeralda Mine and have 3 projects in Goldfield, Nevada. And Converse, the Converse project feasibility, hopefully in a couple of years we'll have those up and running. The plan is to grow the company. And you know, just do like we did in the past.

How are you milling your ore?

In what they call a CIP, which is Carbon-In-Pulp. You have a jaw-crusher which breaks it up. Bring it up from underground, take it through the jaw-crusher, goes through a cone and into what they call a SAG mill. Which is a Semi-Autogenous Mill. From there it goes into two ball mills. It's separated with cyclones. The underflow goes back to the ball mills, the overflow goes into a Nelson Concentrator, which is a gravity circuit, and then the rejects from that go into tankage, a cyanide leach type thing. The cyanide passes through the carbon. The gold is loaded into the carbon, then we take the carbon in, activate it and then heat it up, then run a real strong cyanide solution through the carbon, which pulls the gold back out of the carbon. That cyanide solution goes through electro winning cells and you have plates. Then you take that to a furnace and make a button out of it.

International Mine

The Muckin' Machine's Hittin' the Boss in the Ass

"Mining's pretty much the same. You drill holes, load 'em with dynamite, blow it up, clean it up, put down some track and air line, then do it again... "
~ Chip Huffman

I was born, actually, in Texas. My dad was at a rodeo and my mom was pregnant. I was raised on a ranch southeast of Victor, 'til I was about 14, then we moved to Victor. I went to kindergarten through high school in Cripple Creek. Back in those days grade school was in Victor; high school was in Cripple Creek. Cripple Creek and Victor were rivals, but somewhere along the line that all ended. In 1968, after being to see Margaret Tekavee for about the fifteenth time, she was the judge, I ended up getting drafted. I spent 21 months in Vietnam. Got out of the Navy in 1972. I traveled around for about nine months; I'd bought a motorcycle in Japan. Went to Mexico and Canada, a few other places.

I came back here and went to work for the Ajax. That was for U S Energy. I was just a nipper then. I didn't know anything, so I'd go here and go there, get this and get that. Golden Cycle took over, so I worked for Golden Cycle for a while. Then Texasgulf took over. Then Hecla took over and that was the best deal we ever had. Texasgulf was good, they had fair wages and good benefits. But Hecla had real good benefits and they paid us by the foot. Contract, we got $50 a foot for an eight by eight drift. So we made pretty good money then.

Brock was my partner. He came over from Leadville; he'd worked for Hecla there before. We were working on the 3350 level, driving the South Bobtail, that's the vein we were following. We put in some chutes. Mining's pretty much the same. You drill holes, load 'em with dynamite, blow it up, clean it up, put down some track and air line, then do it again. We had two shifts, day and swing.

But Brock and I got so good that when they started paying contract, $50 a foot, we split up and I got a partner and he got a partner, so we could make more footage. When we worked together, I was right-handed, Brock was left-handed, so it worked out great. We had two drills up in the tunnel all the time. $50 a foot was good money then. We were on a Mine Rescue Team, so we had to go to Leadville once a month, or once every couple of months, and train for that.

What did it mean to be part of that Team?

Well, if somebody was trapped in a mine, you know, it's real difficult to rescue 'em. Because it's dark in the first place, and you fill it with smoke, even with a light you still can't see. So, it's a whole lot of training and working in the dark, and having maps, and blowing smoke around to move it. Putting lines in you'd have to follow because you can't see. We'd go into other mines and teach them what to do in case there was a fire. I was told a mine fire's like a hotel fire, only turned up and stuck down in the ground.

Well then, there was always the thing of high-gradin'. We went through some real nice, real high-grade ore. And you know, Hecla was real good. If you show'd 'em a piece and said, "Can I add this to my collection?" they'd let you have it. But a lot of times when you got off the level at night, if you were working swings, they were up there at night to check your boots and hat and lunch buckets and make sure you're not taking it all home. Hecla had good bosses; they were good people. Then every mine had a Safety Man and Engineers, who would measure your footage and make sure yer goin' the right direction, and your rails are on the right percent of grade. You gotta ditch on the side that runs the water back to the pump station.

You know, the Ajax, I worked on the 31 and 32 for a long time. Gene and I would go from the 2000 level of the Ajax to the 1700 of the Cresson. We were sampling. There were so many tunnels back there, we had to paint our way with spray paint so we could find our way out. I walked two miles underground, everyday. From the Ajax to the Cresson. We were just samplin' old minings to see if there was any gold left. I got to see the Cresson Vug, all the old track and the cars. When they closed the Cresson down everything was still in it. They just left, and never went back. Some parts were collapsed. That's the Roosevelt Tunnel, all the mines hook to it. Comes out on the Shelf Road, and the next one's the Carlton Tunnel, which we had to walk out it many times, too, when the hoist broke down. That's a six and a half mile walk.

What level is the Roosevelt?

The 1700 foot of the Cresson. 2000 of the Ajax, 2000 of the Portland, but they all drain into the Roosevelt.

Could one walk from Cripple Creek to Victor underground?

No, I've never done that. The farthest I've been is all the mines around the Cresson. The Gold Sovereign, the Lame Duck, the Portland, Independence, El Paso. They were all connected to the Roosevelt Tunnel. So, from the Roosevelt you could go in to them. It's true, there are so many tunnels down there you can get lost. In the Vug, where the Cresson found all the gold, that's a huge hole. You know a Wheat Lamp that fits on top of your head, they're pretty powerful lights in the dark. The holes are so big in the Cresson you can't shine one of those lights across it. You can't throw a rock and hit the other side. That was the Vug. I'd have liked to have been there when they done that, but that was a little before my time.

The first time I was underground was scary. It was at the Ajax. I was working for U S Energy. They gave me a light and a hardhat, boots and everything. About two hours later, my light went out. So I had to try to follow somebody back using their light, which is not hard to do, that light moves where ever your head does. Other than that, it was an experience. I was enthused; I couldn't wait to see what was going on down there. I was 22.

What were the hoistmen like?

Well, Bill and Matt were always real smooth, always on the money; never had a problem. You know, they'd leave and come in to the stations, when they'd stop and go different speeds. And then you had some bad ones, you could never tell where they were going to stop ya. Might be five feet above the gate, or five feet below it. Sometimes they'd dunk ya in the water, if there was water below 31. 3100 hundred's the Carlton, below that's a pump. If the pumps go out, it fills up with water real fast. So if it wasn't pumped, power went out, or pumps weren't working, that 250 feet was all water. There was a bell cord inside the shaft, so if you hear the bottom of that skip hit the water you yank on that bell, and he stops it before he fills it full of water.

It never drowned me, but I got my feet wet a couple of times. That's scary, because in that shaft there's no where to go. Little square metal box with gates on it. I mean, you could get out 200 feet underwater, but then you'd have to find a way to get up through all the timbers in the shaft. One skips' goin' up, the other's comin' down, because they're counter-balanced. So, if something hangs it up, it stops. The only thing separating the two compartments is some eight by eights. If it stops, and the cable piles up and gets over in this one, then the other one going up or down catches that cable and it tears everything apart.

Were there pranks underground?

Yep. People eatin' your lunch, or swapin' lunch with ya. We had one guy, John Nothaus, we had big florescent lights on 3100 hundred level, and he grew peppers and avocados, and mushrooms. Because it's always about 90 degrees down there. Everything. Water temperature. Air temperature. Rock temperature. So those things grew good. And everybody had a nickname. Mine was Chopper. Then we had a Junior, and Meat, and Genius, Sleepin' Jesus and Masturbatin' Matthew. We had some really off the wall bosses. Ray Nations was an Indian with a serious drinkin' problem. He was mean, but he could be real funny, too. And Loren, he was a miner from somewhere outta Leadville. He had fingers somewhere the size of these beer bottles. He was a huge man.

I was there when a few people got killed by gas. Carbon dioxide is heavier than oxygen. It just starts pushing the oxygen up, fills up all the space with CO2. It's anywhere underground. If there's a low barometer it brings the gas up. Even in the Mollie. The days they had 'electrical problems' and couldn't take tours, that's usually gas. The barometer is what controls it. And in the Cresson, because the Ajax has ventilation, the only real gas problem was way off back somewhere the fans weren't. Or below 3100 hundred, there wasn't any ventilation. It was forced down. But when we were working back in the Cresson and all those mines, there's no ventilation back there. On days the barometer was low, we couldn't work back there a lot of days. You carried a little lamp, called a safety lamp, and that flame goes out at 18% oxygen. That flame won't burn if there's not enough oxygen.

The old timers used carbides. Before that, they used canaries. We used Mapp fuel. Kind of similar to a lighter, but it's in a glass cylinder and another thing, so you can't break it. You always keep one eye on that lamp. Even in the Ajax where there's ventilation, every team, there's always two of ya, everybody carries a safety lamp. You set it on the ground; it fills from the bottom up.

Would raise your lamp if there was low gas, so you could work higher up?

Well, you can, but company policy when I worked there was if that safety lamp went out on the ground, get outta there. Because, it's probably just gonna get worse.

Did the gas ever get to you?

No. I always watched that safety lamp with one eye. Had a couple of bosses that worked for Texasgulf, from Australia. They were settin' back in the Cresson, and Gene and I were back there, I don't know what they were doin', takin' tours or somethin'. They were executives and they were sitting down on the ground, pecking at the rocks and talking. One of them happened to say something about gas and I looked over and I said, "Speaking of gas, your safety lamp's out. You better get outta there." Well, that's the way it is.

What about the blasting materials?

They're watched pretty closely. You can only go get 'em when you're ready to use 'em. Most of the time we'd use dynamite. At the Mary Nevin, a cheap operation, we had to use spitter cords to start it, we would put a cut in the fuse every inch. Then we had the Ajax, where things were a little more technical. We started with electrics, just hooked them up to the box and pushed the button. Then they got non-Ls, which mean non-electric. They're plastic tubes, you just clip onto a piece of det cord, you know, tie the det cord in a knot, put one cap on it, and they're timed. Zero through fifteen.

Then on 3300 where Brock and I worked, there was a lot of water there. Sometimes we'd drill a seven foot hole, and it'd be squirtin' water six feet down the tunnel. So, to load those, we'd drill another hole right under it. Which ever one was squirting less, we'd put the dynamite in. We had a little plastic thing they called pilgrim hats, that's what they looked like. Pushed 'em into the hole, wedged backwards, to hold the dynamite in so it wouldn't wash out. There's a lot of water down there. It ran down the piss ditch. But you know, you're working from here to the face. You're always wet. A hundred percent humidity, anyway, and 90 degrees. The only thing the rain jackets do is keep the oil from the drills off of ya.

Were you ever in any accidents?

Nope, but I've seen a lot of people get hurt. You know John Weidman? Saw him get his finger cut off one day. They were on the station 31, doing somethin', lifting some barrels up and all those barrels got between the station timber and his finger. Cut it off. Seen several people get things in their eyes. Everybody had cuts and bruises, that's just part of the life. I got my foot run over once by a steel tired tram car. That didn't hurt me too bad; after the swelling went down, I was alright. Steel toed boots, they do help sometimes.

What is a self-rescuer?

Oh, they're terrible things. Well, they're only good for smoke. This little canister you carry on your belt, and it's got a thing that clips your nose, and a mouthpiece like a diver would wear. But the longer you wear it, they get hot. After about thirty minutes they start blistering the inside of your mouth. But if you spit 'em out, then you start breathing smoke, so, you live with the blisters.

Did you ever use one?

Once, but it was only for about ten minutes. We had a wood fire somewhere in the Ajax, it was back in there, but the fans were blowing the smoke right through to us. We got on the skip and got up a couple hundred feet, then the air was clear. They were only good for forty-five minutes to an hour. But you know, there's always a way outta the Ajax, it's the Carlton Tunnel. It's got huge fans down there that pull air in, so if you can't get to the shaft, if you can get to the Carlton Tunnel, there's fire doors you can close; water doors. Big metal doors, for fire or flood, and you can close 'em and that'll stop the smoke and water long enough for you to get outta there. You'd be surprised how fast you can walk six miles. I've done it many times, to get outta the Ajax. When the hoist broke down, electricity went off, that was the only other way outta there.

Did they have their own generators?

No. The hoist ran on 5000 volts. Spools of cable would take up this whole room. Motors as big around as this table. So, I suppose it could be done now, but it would be real expensive generators.

All those times the power was out?

Oh, it wasn't just the power. Any problem. Ya had a choice. You could either start climbing up the shaft, in the manway, or start walking out the Carlton, or sit there and wait for the power to come back on. You know, twenty minutes or so, so we'd just sit and wait.

Did the Ajax get hit by lightning?

Oh, all the time. I had a friend, we were sittin' on the 3100 level, eatin' lunch. There was a phone from the surface to underground, and he was talkin' on the phone to somebody up on the surface. Power came down the phone line and knocked him right off the bench. But when that happens, everybody gets together. Where you unload the skip is called the Station, it's kind of a big room. Everybody comes back there, and all but a couple people turn off their lights. So if you're there for a while, you still have light.

Were your lamp chargers electric?

Um hm, and they had 'em on the surface. Anything you needed, you'd just call up and they'd send it down. Well, the electricity didn't go off that often. Most the time, if you needed anything, you'd get it. And some days we'd have to send down rail, wood, airline.

Did you run water lines?

Yeah, two-inch water line for the drills, for laying the dust. Which wouldn't have been a problem at the Ajax, there's so much water down there anyway. But you still have to have water up to your drill.

Did you ever see carbonated water in the Ajax?

Well, I've seen it bubbling, the water, a lot of times. I don't know why. Yeah. Some of it's real stinky sometimes, when I guess you drill into a pocket that's been there for millions of years.

Where there bugs underground?

Yeah, there's a few bugs. They'd get down there riding the pipes, but they didn't live long. Well, the first couple hundred feet or so, there's bugs. But when you get down to 3000 feet, the bugs don't get down that far, I guess. Unless they'd ride down in a pipe, or a piece of wood or something. I saw very few bugs down there.

Did you blast once a day?

Always at least once, sometimes twice a day. You know, we had good enough ventilation, well, actually when we started out there on 3350, on the Bobtail, we didn't, because we had to drive a raise at 31, put fans to blow air down. So, we had a lot of them, then. Once we got down, if you kept your gloves on, you were alright. If you took your gloves off and handled the dynamite, touched your face, especially around your eyes or your nose, you had a headache real quick. It would burn your skin a little. Yeah, you don't use bare hands with dynamite.

Did the Mary Nevin shaft clog with ice?

Yep, it did. I was a miner so I didn't have to fool with the ice, that was for the nippers. But you know Gene Leaf? He used t' have to go down and chip that, and he'd come out in ice up to here, just blue. He'd have to chip the ice off the guides. That's a job I wouldn't have wanted. The Mary Nevin was only 400 feet deep; it was cold. Drills would freeze up because of the compressed air and the water. That was a pretty miserable job.

Would the water lines down in the mine break?

Yeah, they froze, too. Everything down there froze. I only worked there for about three months. I worked for Jim Young, the old man. He had contracted. Me, and Richie Frettered, and Jim's son worked for him. We were in a stope, taking out a vein. That was alright, I liked the work. We got paid by the foot. Sometimes. The hoist never worked right so we could never get the rock out, so we're sitting there doing nothing until they can move it.

How did you move the rock there?

Chutes. Put it in cars, then tram it out to the shaft and dump it in the skip. The Mary Nevin, they were hand push. Only 700 pound cars. You know, the ones you see around. You always have a downhill grade back to the station, 1%, so when you're goin' down hill, you can ride 'em. You just have to push 'em up hill when they're empty. At the Ajax we had diesel and electric, both. We had five-ton cars and motors to pull 'em. We could move more rock in an hour there, than the Mary Nevin moved in a week.

And graveyard, we had what we called a slusher trench on 33, where you'd dump all the rock that comes outta the tunnels. Graveyard shift, that's all they did was haul all the rock out, so day shift comes back in, they have somewhere to put it. They'd haul the ore and the waste, eleven to seven. You'd tell them with the bell what it is. You separate the ore and the waste. Where you dump it. If you're dumpin' scrap, you skip maybe ten feet in the ditch, and that's where you start dumping ore. So when all the scrap's put in, hauled off, then they can call and tell 'em they're gonna start shipping ore. Because that goes to the mill.

Who decided if it was good or bad?

Oh, the shift boss. Well, you know, Brock and I were following the vein. You have a shift boss that tells everybody what to do and where to go. When we were contract, we didn't really work for him. We worked for ourselves. But they had mine people down there that shipped the rock, ran the pumps, took care of water and electricity, he was in charge of all that.

Did you do your own mucking?

We only had to get it out to the Station. Then company personnel shipped it up. We just had to get it back to the Station. We had tram cars and a mucking machine. The hard part was the rail, sixty pounds a foot, and twenty-one feet long.

Lay 'em on their side, the small part, put 'em back in the rails, them push 'em out as you go, 'til you get all the way out. You gotta put ties in 'em as you go through. When you get all the way out, you turn 'em over, and spike 'em down. Lay some more slide rails. Rail's hard work. It's heavy. You gotta beat all them spikes into that wet wood. That was the hardest part, puttin' the rail down.

Did you timber also?

Yeah, everybody has to timber, because you build chutes. If you drill chutes or whatever. If you're working in a mine, it's double square sets against the rods, and everything's done with timber. In bad ground sometimes, in cavey spots, we'd use chain link fence, rock bolts. Most of it's hard, but there are some soft seams and things that aren't real hard, and they tend to crumble, so you wire those. Or we had metal mats. If it was real bad, we'd use them. A lot of times, we'd do it with 8x8" or 10x10" timbers, and put rail on top, so the rock would have to hit the rail and stop. Hopefully the wood would hold it. The wood petrifies down there, because it's so wet all the time. You can take an 8x8" down there, ten feet long that weighs 150 pounds, and in five years it'll weigh half a ton, because it petrifies in the water. It's always wet down there. Now it's flooded up to the Carlton, up to the 31.

When you worked contract were you guaranteed day's pay?

Well, if we couldn't mine we got, I think it was $10.50 an hour downtime. We didn't have to do anything, they just paid us $10.50 an hour because something happened, like the hoist didn't work, they didn't get all the rock moved and we didn't have anywhere to dump it. We got paid for sitting there waiting for them to get it fixed, so we could go back to work. But you still had to go down there and wait, you couldn't wait on top.

Tell me about the Bobtail vein?

In places, it was very rich. It was unbelievable; I've never seen anything like it. It was probably close to the Cresson Vug, in spots. It had hot spots and bad spots, like any vein. But we did go through some hot spots.

We were the ones movin' it. But we didn't, I mean, we were getting paid by the foot. Sometimes we'd see a nice rock and stop and pick it up, or something like that, you had to wash it off, but when it got out to the Station and the mine people who were getting paid by the hour, they went through it real good. And the shift bosses. One of them in particular would get up on the muck pile, with his little hammer, he'd be up there while the muckin' machine was hittin' him in the ass.

So by the time it hit the surface, it wasn't quite what it started as?

Yeah, because everybody had sacks. They found most of it, they didn't get away with it. But because we were the ones minin' it, and we didn't have time to steal it, if we wanted one all we had to do was say, "I'd like to keep this rock for my collection." and they'd say, "Okay."

Every mine has a Safety Man. That's all he does, run around makin' sure everything's safe. We were doin' something and found a rattlesnake, so we cut it's head off. The Safety Man was named Arnold and he'd always talk funny, so we took this rattlesnake back to him. Next day, he gave everybody a snake bite kit.

Contract minin' we were busy all the time. We didn't have a set lunchtime. We ate lunch whenever we blasted, because you had to blow air in there to get all the smoke and dust out. That's when we'd eat lunch at the station. Then go back and start muckin', wet everything down. Bar it down. Start mucking. Get the rails down, get the air and water hung and start drillin' again. Air and water hang on the side of the tunnel, called the ribs, so as you go you drill holes so you can put in an angle, and just hang the air and water on that. Air and water's not bad, the water's two inch aluminum pipe, doesn't really weigh anything. Air's four inch steel, which isn't bad. But the rail is heavy. You know when you go around a turn, you got a machine called a Jim Crow that hooks on the rails and it's got a thing in the middle and ya twist on, and ya bend it around the corner. But you gotta bend both those rails the same. Then ya got a gauge bar to spread between them to make sure they're the right distance apart. Then when you get to tunnels that go two different directions, ya gotta put switches in.

Would somebody bring all that stuff to you?

Um hm. Well, they put it probably two hundred feet away, so we didn't hit it with rock when we blasted and we had to move it from there. After you finished drilling the holes you had to carry the drills back there a couple hundred feet. The pipe and rail could be a thousand feet away. You know, every once in a while they put in a wide spot, so we could put the rail and ties and sacks of spikes and wedges. We'd put one of those, I think, about every thousand feet. The rock from a blast just flies straight down the tunnel, so if you put a little hole there in the side of the tunnel, you can store everything.

Was your dad a miner?

Yep, I was in the Joe Dandy and the Strong with him when I was five or six years old. He had a lease on the Joe Dandy, an old stope. Actually, he was a rancher, but he worked for Southern Colorado Power, too, and did a little mining on the side of that. I went underground with him on Saturdays and Sundays. I'd shovel what I could. You know at six, seven years old I couldn't do nothing. When I got to be twelve or thirteen, then I was a little more help. He didn't do it for long, after I was probably about fourteen then he didn't work in the mines anymore. He had a good job with Southern Colorado Power; we moved away from the ranch. I guess the economics were better. I suppose he came here thinking, like everybody else, I'm gonna find some gold and get rich. Didn't work.

ABKAP Mining

How to Buy Your First Car

"If the high school kids couldn't hack the mines they had to eat a lot of crow. I didn't want to eat crow and after ten, twelve days I broke in."
~ Richard "Dickie" Grater

I was born April 3, of 1924, in Victor. In a house out on 6th Street. The house I was born in is tore down now. My dad had many jobs. When he first got out of high school he went to work at the Independence Mine, but he was workin' for the Portland. He drove the top car down the A-Level. He operated it into the Portland Mine, and it'd come out at the, they put the waste out at the Independence Mine. That was in 1915. Then he went to WWI. During WWI, my oldest brother, Bert, was born. My father come back after the war, and he was still operatin' that A-Level car.

Then a sort of funny thing happened. There was a couple men come up here. They got a job up there, workin' where my dad was running. And they bought brand-new copper wire; that A-Level car was electrical run, it had a copper line on the overhead of the tunnel. They bought this new copper line to replace the old one and these two guys that come up here from Cañon City, they stole the wire and cut it up and sold it for scrap. Golden Cycle wouldn't replace it, and Dad lost his job there. But, he wasn't out of work long.

He did mostly mining; he was working on the Golden Cycle. He made a little history, in a sense, for the town, for the District. In those days it wasn't mandatory for the men to wear hardhats, either up above, or underground. Well, my dad was always safety conscious, and he wore his.

And it was a good thing, because he got caved in on, and it just crushed that hardhat. Just broke it up, like you'd crack a nutshell. He was in the hospital a long time, but that saved his life. Not too long afterwards they had a mining convention in Denver, and they come up and ask if he'd let 'em have that hat. They took it back and displayed it hopin' the miners... the fact that it saved my dad's life. My dad would'a ended up killed if it wasn't for that.

Where did that happen?

He was workin' on the Golden Cycle Mine. I think it burned down during WWII. It's just a little close to the Theresa Mine, across from Goldfield. And then he worked at several mines. When Roosevelt come in they still had mines workin', but gold was $20 an ounce. Roosevelt come in with the New Deal, and first thing they did was raise gold from $20 to $35. When the Depression started and as time went on, some of the mines began to re-open. The American Eagles re-opened. A man by the name of Bob Bennett, his father was on the Board of Directors or something back east. He made Bob come out here to open up, and he brought a guy with him by the name of Bob McCracklin. And they opened that up.

There's a man named Jim Leman, his father owned the Ford Garage. They had a Ford Agency, too, in the garage. I forget his first name. Anyway, he was never a miner, but he knew of a rich stope down in the Eagles. So, he got the lease on it and he come to my dad and told him, he said, "I'm not a miner, but if you'll mine it, I'll pay a miner to help ya. We'll split the money." The helper would come out of Leman's money.

Well, anyway, my dad knew a miner by the name of Wagner, and he was a good miner. So, they took that pillar out. My dad did a lot of leasin', but that's prob'ly the best money he ever made. Then they wanted him to stay on, and he worked for the company. He hit another rich, real rich ore. But about the second or third time he drilled into it, when he come back the next morning, he drilled into an open stope. Somebody'd got it from the other side.

Did he take you underground when you were a little boy?

He took us on that A-Level. We rode the empty car in. It went all the way from the Independence Mine, through to the Portland Mine. The Portland's right behind and up. They were haulin', shippin' ore outta there. I never went underground 'til a long time later.

Anyway, my dad quit. There was a tractor-man on the county job open, and my mother, because of him gettin' caved in on, she wanted him out of underground, so he took the tractor-man job. By 1936 and '37, gold mining in Victor was doin' real good. Because other miners come back during The Depression. They were startin' to come around, but the rest of the country wasn't.

Before my Senior year, in summer, when school was out, I wanted to have a car for my Senior year of high school. And I knew I couldn't have one with what I made working at the grocery store. Course, Mrs. Hatcher paid me pretty good. Two weeks before school was out, I walked up to the Ajax. I was goin' to run over to the Cresson, and then walk back to town and go to school. They were hirin', but it was underground and parents didn't want their kids to go underground. Well, then finally, the last two days before school ended I took a job at the Cresson. That's when I started underground.

They let you underground even though you were a kid?

I was old enough to go underground; a high schooler. A lot of the high school kids, like I started the summer before I graduated, they'd work at the mines and work on top. But they also, in fact Al Beebe, that was Superintendent of all of 'em, all three of his sons, Henry, John and Evan, when they went to work in the mines, they went underground. They saved jobs for the Shift Boss kids, and they went underground.

I worked in a big stope, where there were four machines. I was a Stope Mucker. And it's muckin' off of the rough. I always thought I was a hard worker, but that day's work was over, I'd take a shower, then I'd come home, eat and go to bed. If the high school kids workin' in the summer couldn't hack the mines, they had to eat a lot of crow. Every day I'd get up to go to work, I'd make myself go. It took about ten or twelve days. I didn't want to eat crow, and after ten, twelve days I broke in, and I just loved it from then on.

What did you do?

They got these machines in these big stopes, and I had to muck their set up. In other words, they'd shoot at night, they blow all that up; they're goin' up. And they load their holes, then they explode 'em. Next day, you come back, then you gotta set up, they got a set up to go higher, but they had to have 7' set up for a stope. So I had to muck their set ups. Well, once I got real good at it, I mean broke in, not good at it, they were easy to muck off, and I'd get through in a hurry. One of the guys, name was Tom, and he is in the beginning states of miner's consumption, from dust. So, I'd hurry. I also had to take care of their steels, deliver their steels to them. I'd get done and Tom would collar the holes, and then I'd finish drillin' 'em. Then they taught me how to collar.

So all that summer, I learned how to run a stope. It give Tom a good break, too. Then when the summer was over, then I had enough money. One of the guys I was workin' with, Rhett Douglas, he just traded in a convertible coupe Model A, in Colorado Springs. He told me, "It's a good one." So that Saturday night, I had my money I'd saved, I got a ride down to Colorado Springs, it was still on the parkin' lot, so I bought it.

The war took a lot of the younger people, and the older people. Gold wasn't essential. It kinda put the District in a bind, but they still kept the big mines goin'. A lot of people left. I graduated in early June, and was inducted into the Navy June 25th, 1943. So, I went to war. Twenty months in the South Pacific at one time. Then when I come back, the war was still on. But half-way across the Pacific comin' back, was when Germany surrendered to Europe. But we all thought that it'd be another year. So I come home on a thirty-day leave, and I was drunk all thirty days. I spent a thousand dollars. My mother finally was so disgusted with me she said she'd be glad when I went back to the South Pacific, where I was half-way safe. I felt bad about it afterwards. I blew all my money.

After getting out of the Navy and a lot of other jobs, I come back to Victor. I got up one morning after a Saturday night drunk and my mother said, "You either get outta the bars and get a job, or you get the hell outta Victor." That's when I went on back to work in the mines. I went up and Bob Welch, who was foreman at the Portland, Salisbury was the

Superintendent, and that's when they were driving the laterals. See, when they put the Carlton Tunnel in and drained all the water, it stopped at the Portland #2 Station. But it didn't drain the Vindicator. And they were runnin' those laterals. So, Mel gave me a job there; I worked there. I really loved that, drivin' those tunnels. But finally it come to where they had to shut everything down to build the mill, and take the railroad up. The last mine they closed down was the Ajax, and they put me to skippin' on the Ajax, until they closed down.

What is skipping?

I would ride the hoist, and then they'd fill it. You'd just go with every station. There were two guys, the other guy was Moon something, a guy from Cripple Creek. But finally they closed it down. Well, they weren't far enough along with the mill to hire guys. I read in the paper that oil shale mines were hirin'. So I and Edgy Cox and Harry Groves went over and put applications in. When we come back, based on the interviews and experience, I thought I had me a job. But I never got it, I never heard from them.

So I went to Bob Welch, he told me when we was still workin' on top that the welders were gonna be the highest paid, and he had Butch there, I can't remember his last name, and he taught me how to weld. Then they called me back to work as a welder on the mill. As it went along it turned out that Kenny Lorenz and I were the high welders. I'd forgot all about that job in Rifle.

Then before Christmas I got a notice, in 1949, that the job was open if I passed a physical. There's enough oil shale between Rifle and Wyoming and Utah, that it would'a lasted this country a hundred years. They could produce it, but it's too high to do it. I went over there and went to work. Just before I went over there they'd put a brand new diesel train in there, an engine, and they fueled it up with the fuel they got from the oil shale rock. See, they had to melt it to get it out. They had retorts. So they were goin' to make a big display, take it to the west coast. Well, before it got to Grand Junction, it burned the engines out.

Later I got sent up to the mine. I liked to mine. You'd drive yer cuts into the side of the mountain there. The tunnel to go into the mine was 30 x 30 feet. The pillars inside the mine were 30 x 30 feet. I really liked it. Bein' as I was helpin' I was told that when they started hirin' again, they'd make me Head Electrician. I don't know quite what happened. I loved workin' in that mine, but the Foreman at the time retired, so they took me down to the plant. Then they hired three or four electricians. After they told me I'd be the first one. I was working with these guys; one of 'em, he didn't know anything. I finally asked him, "What'd you do before you hired in as electrician?" He said, "I was a cabinet maker." It just had me goin'. I was what they called the 'grunt hand.'

Then they put me workin' in the shop. But I figured, the guys I was workin' with, I had more schooling, I was a lot better electrician. They were terrible. I thought, I'm not gonna put up with this. My new car was paid for, I'd saved some money. I went in and told the Foreman I was leavin', he looked at me real funny, cause he and I were pretty good friends. I think he'd forgot all about me. I went to Pueblo and got a job with Southern Colorado Power Company.

Later I went back to Rifle, and I worked at the Uranium/Vanadium Mill. The U.S.S. It was vanadium only, 'til that war come along. Then they needed uranium for the atomic bomb. I did work in the Plant where they broke it down to what they called yellow cake. See, the vanadium company made millions because they'd been minin' vanadium so long, and then the tailings were right close. Well, a lot of times uranium and vanadium go hand in hand. And then when they needed uranium, well all they had to do was put an acid leach in, and I don't know how many years it took 'em to process those tailings. But that was a gift to 'em. It wasn't dangerous stuff, at least not in the mill, I don't think. But the miners, some of the deaths later on was contributed to mining, but I can't vouch for that. I think we left in October of 1954, and I went to California. I went to work at Douglas Aircraft. It's Douglas McDonald now. I retired and moved back here in '82.

Mary Nevin Ore House

Every Job But Boss

"I've run machines. I've laid track. I've run loaders.
I've done everything there is to be done underground." ~ *Willie Roy*

Elkton was the first mine. I was 24. I've run machines. I've laid track. I've run loaders. I've done everything there is to be done underground.

What kinds of jobs are there in the mines that you haven't done?

Well, being the boss, that's one. (laughter)

I worked on some levels where there were maybe twenty-four guys, and that's about the most. At the Ajax where I worked, that was the most men. And then again, I'd work where there was only two. I think the lowest is two; there has to be two.

Is there any danger of miner's consumption today?

Well, yeah, there is, if you don't... there are a lot of people tramming, that's taking the muck out of the chutes and the stopes, take it up to the station and that has a lot of dust, and you get it from there. It is as bad, almost as bad, as the machine. Excepting the machine, when you hit the back. When you hit that back, all that real dust, and in just a minute or two you can't see nothing. Wet machines, you don't have that problem.

Can you think of any other equipment that isn't the same today as it was in the 1930s?

Well, yeah, I'll say a Leyner is something you set up with a cross-bar. You use the side of the stope, one side to the other, and you put your machine on that, you got a cross-bar. Well, you don't have that these days. You gotta be a machinist to run the drill, that's your difference. The machinery is all modern now. Stopers are so much lighter. So much lighter, it's just easier mining. When I first went to mining, it was all hard mining, you know, it was all harder. Everything the hard way. Which it 'aint now.

What are the temperatures at the different mines?

Your working conditions in the winter were no different than the summer. Your temperature down the hole was always the same. Winter or summer. No different. I don't know just how warm it would be, but it's a lot warmer than it is in here. You had to shed all the clothes that you could to work. They had some freezing in Elkton; they had to chop the ice out to go down the skip. But that was mostly up in this North Stratton cut, not down in this part by the Ajax. You have one mine hooked up with another, and there's a draft going through. The Ajax has a draft down. That's why a lot of the mines are gassy, there 'aint no air circulating.

It gets warmer as you go down. That's why on your lower levels, say thirty-one, well, not anymore, they're all connected up with the Portland, Rose Nichols, all of these mines are hooked together. But back in the drift, well, that's different. Then it's hot 'cause there's no air going in it. I mean at several hundred feet.

We had to work with what they called a 'pressure door.' A pressure door you blocked off your grid, with that door. Then you started up some air pumps and pushed air back into there. It's the only way you can work, when you get one of your stopes through. Because every hundred feet it's a stope, that's why your levels run every hundred feet. When you hooked up with another stope, then you'd try going through.

What were the different depths of the different mines you worked?

Some were 600 feet deep. Some was less, 300. The Cresson was 1700.

Tell me about high-grading?

Well, your high-graders here in the early days, there was always somebody would buy your high-grade. Now who he was, we wouldn't know. Anyway, I'm going to use one person, he's not here, his name was Bill Moore and he was an assayer. So he bought the 'high.' Nearly everybody had somebody to take their high to. Your high-grade depended on how good it was. They could've paid you by the pound. You could say that if you had high-grade that was only three dollars a pound, you didn't bring it home. It had to be better than that. If you'd get and could carry three pounds, three to eight pounds of rock in your bucket, and bring in twenty or thirty dollars, you would leave it down the hole. That was your high-graders.

Before my times here, they'd run it up in their hair. They pulverized it. They took the rock and pulverized it, they were that good. They put it in their hair; their buckets were searched. But in later years, why, they didn't have that, so they'd fill up their buckets.

When you went down at the mine you set your bucket down. When you hit the surface, somebody took your bucket from there. If your mines were pretty rich, well then you went in the change room your clothes was up here, hangin' up, you couldn't reach 'em. So then when you took your other clothes off, they'd let you have them down. Then you could change and your other clothes were checked over. But that is how they got away with it in the early days. It always seems funny, but a lot of them swallowed it. So don't try that, I never did.

Were there still railroads, or electric lines when you came here in the '30s?

Yeah, when I came here the railroads was all here. They didn't go out until around '48. The Trading and Transfer would haul a lot of it down below. There weren't no street cars then.

I was told to ask you about the Trading and Transfer?

Well, the Trading and Transfer, all these little mines, they hauled their ore to the mill. Well, they hauled it, put it in railroad cars. They had a fleet of trucks. All the coal that was used up here, the Trading and Transfer

Company brought up, by railroad and by trucks. They had everything. Hay, grain, all your groceries, sugar and everything was hauled by trucks.

What was the most salary that you made?

The most salary was a dollar and a quarter an hour, which made you up to twelve dollars a day. Eight hours a day. I worked a lot more than that, because I repaired all the shafts and any extra work I usually got.

Weren't you allowed at that time to do some leasing on the mine and work at night for yourself... how long did you work at night?

You were supposed to work at least eight hours, but you could work four hours and then come up. You had to come up when the bosses came up, when they came up you had, you know, four hours. You had four hours, that's how long they worked. Four, usually four, if we were leasing. Then sometimes you'd work for the company, it depends on what your breakdown was. That's why you needed more hours. Ya done better than working for wages. A lot better and not work as hard.

Your work in the mines is all dangerous. If you're running a motor, you've got to be awfully careful, you could get hooked. If you're running a machine you've got to watch that, watch the back, so the motors don't come down and hit ya. There's been a lot of guys that got hurt that way. You make your safety by you watching. Being careful.

Were there ever any cave ins?

Yeah, there were several. The Cresson had a big one when they lost their stope. They went there for several months filling that hole, and they can't mine it again. They pretty near lost everything, when they lost everything at the Cresson. There is a lot of small cave-ins, that was one of the biggest.

Mary Nevin

Charmed, I'm Sure

I heard them comin' back, they banged on the airline or something. And when they got there I said, "Don't pull the chute, I'm in it!" ~ *Ted Johnson*

I was born in Chicago, 7-10-49. Miner forty-niner. Born one hundred years after the California Gold Rush.

How did you come to be in Cripple Creek?

Oh, my cousin, Jim Brownlee and I came on our way through to the West Coast. We came through Colorado. First time I saw the mountains I said, "This is it. I want to live in the mountains."

I had a good job, workin' at Western Electric back there, had a new car and a motorcycle. I told Mom and Dad, "Oh, I'm movin' to Colorado. To the mountains." So, we'd just gotten back from vacation and Mom and Dad were just leavin' on their vacation. They were gone for a couple of weeks, in the mean time I sold my car and my motorcycle. Bought an old '63 International pick-up to come out here. So they got home, saw this old truck in the driveway, they said, "What's that hunk of junk doin' here?" I said, "I'm goin' to Colorado." They said, "You won't make it to the State Line in that thing."

My cousin and a friend of his were plannin' on getting the old Evergreen Station and making a woodworking shop. I was going to work with them. Couple weeks before I was goin' to move to Colorado he called and said,

"Well, the deal fell through, we're not going to get the place. I don't have a job for you." I said, "Well, I'm comin' out anyway." I just loaded up all my stuff and came out, and I stayed with some of my friends until I found my own place. After a week, I got a job at Cliff's District Service on the corner where the Gold Rush Casino is now. You know that big ol' gas station sign? Well, that used to be a gas station, and I used to work there. 1973.

I worked there, then winter came and he laid me off. I got a job working for Ernie Snare out at Mountain Estates, clearing roads. People would go through and cut the trees, then I'd come through and stack it to the side, then they'd come through with bulldozers and cut roads. And then after that, I went up to the Ajax. I needed a job. I heard: "They're hirin' at the Ajax." I didn't even know where it was.

So I went up there. It was actually kinda spooky goin' by the Portland I and the Portland II. It was dusk, I'd never been up there before, and there's the Ajax over there. It was just kinda weird. So, I went in and talked to the Superintendent, Louis Yuturby was there at that time. I got a job. The first couple of months, I worked on the surface, tearin' down the old ore house. And I read books about the area, I was really fascinated by the history. And I'd see these guys goin' down the hole every day. Stuff comin' up, we're sendin' stuff down. I was really curious. I told Louis, "I want to go underground. I want to go underground." He said, "We'll see, we'll see."

After a couple of months, we were in the lamp room, clockin' in and Louis said, "Ya wanna go underground?" I said, "Yeah!" "Well, one of the samplin' crew didn't show up today and we need somebody to go down." "Okay!" So what they'd do is go down on 2000 of the Ajax, go on back to the Roosevelt Tunnel and back to the Cresson, and get samples.

So, here I am, I'm all dressed up to work on the surface. Insulated boots, long-handles, you know, all bundled up to work on the surface. Then they ship me down this hole, 100% humidity and 80 degrees. I went walkin' back and I was hotter than hell. You know. We finally get back there, then we fill up sixty pound sample sacks, gotta pack on our backs. You talk about sweatin'. I was just soaked, beat to death after that first day. But, I loved it. I got down there and I said, "This is for me." I just loved it.

Then for the next day I said, "Well, am I gonna work on the surface or underground? I about died down there today with all these clothes." He said, "Oh, just figure on goin' underground." "Okay." So, that's how I got started minin'.

What it is, they had a crew at the Carlton Tunnel raising track. See, I think there was two or three layers of track in there and it was underwater. And there were some places the sand would wash up and you couldn't get the motor through. So, I worked there a couple months, just raisin' track. Jack it up, put rocks under it.

After I did that for a while, Old Man Young, Old Jim Young, he was a shifter or something. Anyway, they stuck me and my partner back in the New Market, driving drift. That's the first drift I ever drove in my life. Old Jim would come in, and he'd take spray paint, mark where he wanted the holes. We'd just drill out where he had marked, then load it, shoot it, muck it out. I loved it, oh, it was great. I loved runnin' the muckin' machine and drillin'.

One day I was back there, this muckin' machine has a big pin with a big nut on it. You got your travel motors, forward and backward, down here. And the mucker box sits on top of that pin. Well, the pin is supposed to have a cotter pin in it. I don't know if it was broke off or wasn't in there, but anyway, it kept working loose. And I'd notice the machine would be rockin' when I'd go to pick up a dipper of muck. And the box would tilt forward. And when you'd go to fill up the car, it'd tilt back. Then finally the nut came off, and I had a load of muck in the dipper. I tipped that handle to raise the dipper, but the dipper stayed there and the box lifted up off the carriage. Then it started tippin' over on me.

There wasn't much room in there, got this handle on the side and the box was comin' up, tippin' over. I'm trying to get the handle to where it'd stop it, because it was comin' up. And it was just about ready to hit my head, crush my head into the rib, but I got it stopped. There were a couple guys there that pulled the thing off and took me out to the Station. They sent me to a doctor and I had three broken ribs. I was back the next day. Couldn't do anything, but I was there.

The first time I saw a muckin' machine, it was a noisy grunting son-of-a-gun thing, you know. I said, "Boy, you'd never get me on one of those." But down the line, I got real good at mucking.

Who taught you to mine?

It was old Jim Young. He was, like I say, a shifter. He'd get it marked out, tell us this and that. Come in and look at it, show us how to load. My partner was a green hand, just like me. We just had Jim to keep an eye on us and show us what to do.

I worked there four or five different times. I'd work six months, get laid off a couple months, go back for a few months, get laid off. So, I guess I got laid off and the next time I went back U.S. Energy had the contract sinking shaft down to the 3400. So they'd be shipping shaft muck, and we'd have to, a lot of times we'd haul it down the Carlton Tunnel and just dump it in the piss ditch, you know. Goin' in and out of that tunnel, you were always derailing the motors. We had five-ton side-dump cars and they'd derail. Had to jack them back on the track.

One time when we were sampling in the Cresson, we put in our shift and were comin' back to the shaft. We were almost there and decided to sit down and take a break. There were four of us. We were sittin' there having a cigarette and I heard something. "Do you hear that?" "I didn't hear nothin'." I mean, it's dead quiet down there, you know. When you get away from the machines.

"I heard a noise. It sounds like somebody's talking." "Yeah, I heard that. It's coming from where we just come from." Who knows how long it'd been since anybody'd been back there, except us. We're sitting there listening and we don't know what to think. It's getting louder, and louder. It's two guys talking. We don't know what in the hell is goin' on. Turns out it's Jack Ham and Al Tapp. What happened is there was a wreck in the Ajax shaft about, I don't know, 700 or so. Anyway, they had a wreck and one of the skips came out. See, they're counter-balanced. One's comin' up while one's goin' down. So one of them's goin' down, it bounces and dogs off. The hoistman didn't notice, so he kept dropping cable into the other compartment. The other skip's comin' up, like 1000 feet a minute, hits all this cable, just tore the hell out of the shaft. But we didn't know it at the time.

So what had happened, they got these two guys, they were engineers and geologists at the mill, Jack and Al, and they went down the El Paso, which hadn't worked in years. They walked over toward the Ajax, they had the maps, got us. We went over to the Ajax, couldn't see anything, so we walked out the El Paso. Ol' Dale Weaver had it at that time. He is something else. A few weeks before this happened, he was workin' on a CAT. Somehow the blade dropped and he broke both his feet.

So, we got down to the El Paso shaft and it was like quicksand down there. I mean, ya got your mine boots on, but you'd just about walk out of your mine boots. It was like quicksand. So anyway, we finally get to the skip, which was the old fashioned kind where you put the car on, and you take the whole car up. So we finally get on the thing, then we got 1300 feet of cord. So I'm pullin' on this bell cord and it goes about three feet before it feels like you got the signal. So I rang 3-1. Nothing happened. I rang 3-1 again, nothing happened. I rang a 3-1 again, nothing happened. I rang a 3-1 again, finally, he starts pulling us out real slow. I think it's safe to say that was the last man-trip out of the bottom level of the El Paso. We got to the surface, and he comes hobbling out of the hoisthouse with two casts on his feet. That was before it burned. Well, then it did burn, and it messed up the headframe too bad, and the collar. Then down the line the collar caved in.

Dale knew to expect you because the other guys had come through there; they couldn't just call?

Yeah, he let them down. There was no phone back there so they came and got us on the 2000 level. The guys on the 31, they went out the Carlton. See, back then the manway only went down about a thousand feet. So there wasn't even a manway down at the Ajax. That's why they had to go down the El Paso, come over to the Cresson and take us back to the El Paso.

Was it an adventure?

Oh yeah. You get back there in the Roosevelt Tunnel, see the stalls and water troughs where they used to keep the donkeys. Just unbelievable workings when you get back in there. I was just fascinated to see what all the old timers had done. The workings in some of these mines is just incredible. Huge timbers, you wonder how they ever got them down there and in place. The old tram cars down there, all kinds of old equipment. I appreciated what these guys were goin' through to get this all done. Some guys mentioned the 1704 Stope. That thing is so huge, from what I understand it is an extension of the Cresson Vug, which they hit on the 12th Level. 2000 Ajax is 17 on the Cresson. And you'd be walkin' down a drift and all the sudden there's nothing. You know, there's no board across or nothing. You'd shine your light, look down, look up. You can't see anything; it's just unbelievable.

Back when I was runnin' the Mary Nevin, they used to take geologists back in there to what they always called Gold Run. There's Old Gold, Henry Adney, C K & N, El Paso. We'd go around to all those mines just looking and takin' samples. One time, going back we came to this big, huge bowl. I thought I was lost for a minute. So there was a little sign on the shaft, and it says "Flash 1-3 for Cage." So I said, Okay, we've got to be on the 3rd Level of the El Paso and the whole thing caved in. I mean, this huge three-compartment shaft, 10x10, 12x12 timbers is just a masterpiece. And one day we came back there and I mean, it's just all at the bottom.

So then I got on this old rusty airline that goes across the shaft to get the sign that says, there it is (pointing to his wall) "Flash 3-1 for Cage." I was going across this old airline, across the shaft, and this geologist says, "You're nuts. Leave the sign there. You're nuts." So anyway, I went across there and I got the signal chart and the cage sign. I could'a gone down the shaft real easy. But I was young, foolish, you know.

Being laid off from the Ajax, I guess then I went to work at the Gold Bond. I was working at the other gas station in town, Marsh's. Al Goodwin would come in and get gas and stuff. He was startin' up this mine and I said, "Hey, you know I want to go to work." "Okay, it'll be a couple weeks." So I went over to the Gold Bond and it had an old V-8 Ford engine for the hoist. We were retimbering the shaft at the time. Once we got the shaft done, we started driving the north drift on the 400 level.

One day, well, Brian Marshall was the top-hand. Greg Pratt ran the hoist. And it was me, Wild Bill Haley and Nervous Nutall in the hole. And one day, I came in to get mucked out and there was this huge slab. Hey, I'm going to give Brian a little surprise. So I tried to get this big, huge slab into the dipper in the mucker, and it wouldn't get in. Finally, I just got it teetering in there, then I had to back up about 20 feet where the slab would clear the back and fit in the ore car. So I finally get it to the ore car, take it out to the shaft. It was so big, I lost the car, it fell into the skip, and then I had to wrastle around to get the car out, then I finally sent that slab up.

On the surface, the ore chute, there's a lip that sticks out over the ore car, so this slab comes up, goes down that chute, hits the ore car, knocks the ore car over, and when the ore car tipped over it pulled up all the planking for the chute. Really messed it up. Scared the shit outta Brian, he didn't know what was up. Anyway, a few minutes later the phone rings and Al's on the phone. "I don't' ever want to see a rock that big come outta this mine again. You tore up the chute. It'll be at least an hour before we can start shipping again." You had to be there. I had no idea it would cause all that damage, I just thought they were goin' to see this great, big slab comin' up and be surprised. And he was.

Al didn't have much money, so it was 4x4 headframe. One time we were shipping muck; it was the first trip of the day and we got the car loaded up. Sent the skip up and it didn't come back. And the phone rings, it was Greg. He says, "Al says to climb out." I said, "What's the matter, what's goin' on?" "Al says to climb out." "Okay." So I'm thinkin', well, maybe he's got a mine inspector there wanting to see if the manway's passable.

So I'm the first one up, stick my head out of the manway, here's the sheave wheel and a big pile of timber laying on the ground. The skip's dogged off the collar with a ton of muck. Al's sitting on the bench by the hoisthouse with his head in his hands. What had happened, is when ol' Nervous Nutall had... see, there's a safety chain on there. So when you got a mantrip, you have the chain on so the skip can't tip. When you start shipping muck, you take the chain off. It was laying on one of the rollers, you know the rollers on the back of the skip come up, hit the scrolls to dump it. Well, the safety chain was on the roller, came up and hit the scrolls, and it pulled the sheave wheel out. The sheave wheel came down, the dogs kicked in, parked it on the collar. (laughter) I stuck my head out, saw that pile of timber and that sheave wheel, Al with his head in his hands. We had to fix the old headframe; just took the bad timber off, saved what's good, rebuilt it. With 4x4s again.

It was just one of those things. If he had been a little more careful and made sure that safety chain was hanging down and not laying on the roller... it was just a freak thing. You do it every day, and this one day it happened.

Then I started at the Mary Nevin in '76, and it closed about '82. George Shrank had it back then, and somebody told my wife they might be hiring at the Mary Nevin. I went down there and talked to him, and Mike Altman and Emery Carter were cleaning the shaft. And Mike would quite often dump shift, get drunk or something, or just not show up. So George hired me. So, me and Emery. He had a little one-cylinder gasoline hoist that we used for bringing timber up, and lowering timber down. Oh, sometimes that thing wouldn't start. You'd pull start it, and sometimes it wouldn't start. A lot of trouble with that.

Oh, I loved it. A lot of places caved in, so you're just workin' off the muck. And the old crib timber, it was all caved in. So you had to take this cable, wrap it around the timber and pull it up with this little tugger. And for the most part, it was frozen in the muck and you'd have to take a hand-steel, drill down six inches or whatever, and shoot it to break loose. One time I was sending a timber, and somehow it got hung up in the shaft. Emery said, "Come on up." I went on up. And when that timber had lodged in the shaft, see this little tugger had a 4x4 from the base of the tugger up to a timber in the headframe, to keep it. That's how we had it mounted on the ground. And when that timber hung up, it broke that other timber, and the tugger started climbing up the headframe. And finally hit that timber, and the engine died. That's when I came up, and see here's the tugger, half-way up the headframe. It took about a year all together, but we finally got down to 410. Then we started to do some minin' on 270, 350.

Well, then there's the time when I got buried. It's mentioned in the book a couple of times, but see, it happened on a Thursday afternoon, and there were reporters out there from the Gold Rush when I got out about 3 o'clock. The paper went out the next day, so they never even interviewed me. They just had second hand information as to what was actually goin' on down there. If you want the whole story, I can give it to ya.

They were pulling the chute on 410, and there's an old stope between 350 and 410 that we used as an ore pocket. So we'd dump ore into that and pull chute on 410. Well, if the chute started gettin' empty, it'd cone out. I'd go up, on a kinda like sublevel, and it's a big area that's funneling down to this big ore pass. So I'd throw a few rocks, and start breaking that cone, and we'd get another fifteen cars from around the edges. Well this one day, I'd done it a hundred times, but this one day I got too close to the edge and it gave out from under me. I was floating on the rock, heading down into this ore chute. I had my hands over my hardhat to keep it on. As I was goin' in, I could see all the other rock around the edges comin' in behind me.

So I got down into the chute, and I had all this muck just keep runnin' up above me. And I thought, holy shit, I'm dead. I knew the guys were gonna be comin' back, start pulling the chute. I didn't know if they were goin' to be able to hear me or anything. You talk about a sickening feeling. You're on an incline, my left arm was out like this and it was buried up to my shoulder. So I couldn't move it. I was right up against the back; right up against the rib. I had a little space, like here's the rib, and I had just a little air pocket in front of my face. I forget who the guys were pulling chute. Bill Snare was one of them. Anyway, I heard them comin' back, they banged on the airline or something. And when they got there I said, "Don't pull the chute, I'm in it."

I found out later, they had the fire departments there. Everybody in the mine was there, and they started digging me out. From above. So you hear the sound of muffled voices and shovels. You can't imagine. They got down a few feet, and I could see behind me that there was an old loading stick. They said, "Hey, we see this loading stick, where are you?" I'm trying to envision up there, where I'm at from the sounds. I said, "Move over five feet towards the rib and start digging there." So, they moved over and started digging there. They finally got a little hole opened up to me. I thought geeze, I think I'm going to get out of here after all. They tried to get an oxygen bottle in there and it wouldn't fit. All the sudden I hear all this noise, and the whole thing caved in. See, they had to timber as they went down to support the rock. Well, that let loose somehow, and when it caved in, the muck ran in and filled up the little space I had by my face. I thought, well, that's it. I'm really dead now. So, they kept diggin'.

My arm, my left arm that was buried, it was numb. I thought if I do get out, probably they will cut it off. But I wasn't in any great pain. Then they finally got a hole opened up to me again. I passed up my hardhat to 'em. I had to use one hand to undo my mine belt, with the battery and self-rescuer and all that. Passed it out. And slipped my arms up. They're trying to pull, they were about to pull my arms out of their sockets. Wouldn't budge. I said, "Okay, you're going to have to start pulling the chute. Real slow." That's when they started pulling the chute, and I could feel the muck finally getting loose by my feet, up to my knees, it got up to my waist. I could feel the pressure comin' off. And about to my chest, I said, "Try it again." And boy they just yanked at my arms, and I came out.

I was fine the whole time I was in there, tell them where to dig, do this, do that. As soon as I got out, that's when I went into shock. My legs turned to rubber, I just collapsed. They carried me out and took me down to the Springs to the hospital to check me out. I was bruised from head to foot; didn't break anything. But I found out later (we didn't have a phone at the mine) that Bob Phillips got in the company truck, to go into town to get the fire department. Well, he had a blow-out on the way there. So, he pulled over and flagged down some tourists goin' to Victor. "We got a miner buried at the Mary Nevin, I gotta get in to Cripple Creek." So the tourists gave him a ride to town, got the fire department.

One of my brothers was out on vacation at the time. He had my motorcycle. Well, they'd gone into town and got my wife, she was a librarian. Brought her out, she went down to the hospital and everything. But while all this is goin' on, my brother was in Trego's Emporium, and somebody came in, I don't know who it was, "There's a miner trapped at the Mary Nevin." And the clerk says, "Who is it?" "It's Ted Johnson." My brother hears this, so he runs out, jumps on the motorcycle and heads to the mine. By then they had taken me to the hospital. That was just a couple funny things that were goin' on.

I don't really remember much. Two things I do remember thinking about, was I wonder what the last minute's gonna be like, and I kept thinking, Mom and Dad are going to have a fit when they find out I got killed down there.

Over at the Mary Nevin, it's a downdraft shaft. So once you got down around, oh, 250, the shaft would start icing up. A couple guys would get in there and start chopping ice. Misery, oh God. Everybody took turns. What we'd do is we'd go down as far as we could, and the ice was all around the guides and you'd get stuck. So you had to get out in the manway, which is all full of ice, and reach underneath the skip, chop the ice off the guides. That's pretty dangerous, so I had this idea, which was illegal as hell, but we had a forty-gallon drum, and I put cables to it, and clamped it underneath the skip so you could ride down in the bucket and not have to get out in the manway to chop. It was illegal, because there were no dogs on this bucket. We did use a safety line hooked to the bottom of the skip, but if the clamps came loose, you're gone. And if the skip got away, it'd hit the bottom and you'd be crushed. But it was the easiest, the fastest way to get through the ice.

So there'd be a couple guys up in the skip choppin', and you underneath in this drum chopping, after a half-hour or forty-five minutes, the drum's full of ice. Ice is coming down all over the place, you got a downdraft that's colder than hell. So you'd last maybe forty-five minutes, then come up and a few other guys would get in and they'd chop as far as they could. That was just miserable. You get the guides clean enough to where you can get through, but then there's still all the ice up above. You want all that chopped out, otherwise it builds up. So the more you get out, the longer it takes to build up. It was miserable standing up to your waist in ice. We tried hooking up a heater, a blowing fan kind of heater with vent pipe going down the shaft to heat it up. And it helped, it would melt it. It helped, but we still had to do something about chopping ice.

There was this one time, we were almost done with the shaft, we were putting in the Station on 410. We had the shaft down about twenty feet from the bottom. It was a Monday, Brad Place was President and he wasn't there very often, but he was there to see how the shaft was goin'. So we ran the cage, you know with a safety light. The ice wasn't bad in the shaft, so the skip made it up. Well, me and Brad got in and we're goin' down. It's not a real fast hoist, but the bell cord would freeze up against the timber. So you'd have to pull on it and break it loose as you went down. We got down, I don't know about 350, something like that, but the ice was so thick, I couldn't break the bell cord loose. I pulled on it, it wouldn't break loose. So here we're heading down, we got no bell cord. Lawrence was runnin' hoist, and he had a mark where the bottom timbers were. The only part where we had guides. Brad says, "What do we do now?" And I says, "Well, Lawrence has got it marked, he's gonna stop us." So we're goin' down, didn't stop. Ran us down off the guides, skip starts spinning around, we hit the bottom and the skip tips over. Brad's eyes were about like saucers. I got on the phone, I said, "What the hell are you doin' up there? Why didn't you stop us?"

"I didn't get a bell." "Bell cords froze, you got marks." So anyway, Brad says, "What are we going to do?"

So now, the only thing I can do is climb out. I got to climb out, get an ax, climb back down, chop the bell cord loose. I forget where it was, there was one section of the shaft that the manway was completely frozen over. I can't remember how I let that happen, that's the only way out. But anyway, I got up to where the manway's plugged, I had to get out and get into the shaft compartment and you have ¾" hanging rods, to hang your sets, and they're coated with ice, must have been 4" around. So you can't even get your hand on it to hold it, you've got to put your whole arm through the hanging rod, to your elbow, to get a decent grip. It was about twenty, twenty-five feet I had to climb in the shaft compartment. No safety line, one slip and away you go. I finally got up to where the manway was passable, got in, came up, took a little break, and started climbing back down with an ax. Got down to where the manway was plugged, had to get out in the shaft. So you're hanging on with one arm, chopping ice with the other, and finally, I got the bell cord loose, got past the plug in the manway, got back in the manway, got back down to the bottom level.

Brad waited for you at the bottom?

What else is he gonna do? (laughter) I finally get down there and you talk about hair-raising. Whew. Anyway, I get back down there, we got a bell cord now. So I ring the skip back up onto the guides. We get back in the skip, came out, and I noticed that the rollers for the scrolls were on the front of the skip, instead of the back. When we got the skip back up, it had spun around, so it was on backwards. We had to go back down, take it down back off the timbers, spin it around, put it back up on the timber, and we were finally ready to go. So I forget exactly, but I'm sure I had everybody in the manway getting that cleaned out.

There were a couple of times my folks came out. They didn't like me mining, it was dangerous and all that. They came out to the mine one day, got all suited up, diggers, hardhat, lamp and everything, and we went down to 350 level. It was cold, dripping water, you know. Noisy, the guys were drilling down on 410. I told them, "See, it's not as bad as you think." Dad looks around and says, "It's not so bad." Mom said, "It's worse than I ever imagined."

I think it was the next day, Mom and Dad came out to the mine again. Mom didn't want to go down. I said, "Dad, come down and we'll go to 410. Shoot a round." We had a cross-cut back in there, about thirty feet, and they had the round drilled out. And he watched them load. We were shooting with igniter cord, you wire the holes in the order you want to go And so they get all loaded and everything. Light the igniter cord, we can see it, it's like a sparkler kind of thing, spits out sparks. You can see it burning, not like fuse. Once they lit it, we went to get in the skip. Me and Dad are in the skip and the other guy, I forget who it is. And ol' Nervous Nutall, he was the one that spit the igniter. Well, he comes walkin' back to the skip and just as he's about to get back in, his dinner bucket opens up. Sandwich falls out, cookies, you got an apple rolling around. Here's my dad, looking at 100 pounds of dynamite gettin' ready to go off, here this guy is chasing his apple around the Station. Finally he gets it all gathered up, gets on the skip, we got out in plenty of time, you know. That's one of my Dad's favorite stories, "They had the round all wired up, it was lit, ready to go off, and this guy's chasin' his lunch around the floor."

When we got to the surface, we had a piece of plywood over the manway, so if it snows it doesn't go in the manway. So we got up and the round started going off. Well, that concussion comes up the shaft. And here's this piece of plywood flopping around. Mom said, "What's that?" "They're blasting down there. Plywood's flopping around every time a shot goes off." When you're shooting that close to the shaft, most of the concussion is going up the shaft.

When we were drivin' drift on the Gold Bond, we'd shoot on 400. We'd go up to 300 to listen to the shots, because as best as you can, you want to count the shots to make sure all the holes go off. So you'd go up there, and the concussion, if you were smokin' a cigarette, a shot would go off and that smoke would move about a foot away from your cigarette, and then come right back. Every time a shot went off, the smoke would go and then come back. About blow your hardhat off, almost. It'd just go right through you.

Have you been to the Gold Bond in recent years?

I was back there about six months ago. Who tore all that stuff down, do you know? Because for years, I'd go out there, the hoist house and everything was still there. See that "Men working in shaft, don't run hoist" sign, this board here? (motioning) Greg made that, and when we were doin' something in the shaft he'd hang that on the hoist so nobody would move it. I went out there from time to time. Everything was still there. I found that sign layin' there, a few other odds and ends. But I hadn't been there in years, then recently, about six months ago, I went up there. All that's left is a concrete pad where the hoist was. Building's gone, headframe gone, compressors gone. There's nothing there except a concrete slab. There's a place next to the shaft that had been stoped out up to the surface years ago, but now there's a place between where the hoisthouse was and the shaft, there's a big spot that's caved in. But on the concrete slab it says, Reeves Minerals, 1976, scratched in wet cement. That's about all that's left of that.

Did the Mary Nevin make money?

No, when we first started we had a 25-ton a day ball mill. And then Brad came in, we went public with the company, sold stock and everything, and Brad wanted to put in a 100-ton a day ball mill. Well, there's no way that mine could ship 100 tons a day. Just got the one skip, it's not even a 3-compartment shaft so you only have one skip instead of two. We didn't have the ore. There was times, somebody, an investor would be comin' out to take a look around, you know. And I would have to ship waste to keep the mill goin' so that these investors could see the mill runnin'. I mean, I had to break everything I could drill a hole in to try and keep that mill goin'. We had some good ore, some real low grade. You know, we just didn't have the ore to meet the capacity to keep a 100-ton a day ball mill goin'.

We had a line comin' over from the Shelf Road for water, and I remember they were always havin' trouble with that thing freezing. We had a pump station there from the creek. It came all the way over the hill, to the mill. That would freeze up once in a while. Once you got to the mill, it was pretty warm because you've got all those chemicals, flotation tanks. Now if we'd have kept that 25-ton mill, well we could have got ahead of the game once in a while and stockpiled it, so if we ran out of good ore or something, we'd have something to keep the mill going on. But we couldn't stockpile; it came outta the mine and into the mill. We were runnin' waste rock through the mill. Tearin' up equipment and payin' the mill hands, and you're not gettin' anything. We were runnin' 3 shifts a day. I think we had 30, 35 miners all together.

Did you ever run in to bad air?

Oh yeah. Well, Paul & A.J. were killed at the Ajax. They were workin' for U.S. Energy, and I was workin' for Golden Cycle. They rotated every two weeks, we rotated every week. So I went off shift Friday night, swing-shift. I was off Saturday and Sunday; they got killed the next day. In the gas. And I didn't find out about it until Sunday. It was a real shock; I couldn't believe it. I just worked with them, now they're dead?

I went to work Monday, mine inspectors all over the place. The mine was shut down for a week. I believe it was the following Sunday, me and my partner were the first ones down after they reopened it. When 33 flooded, timber floated up the shaft. Then when they got it pumped out, the timber was lodged, crisscrossing the shaft. We ran a safety lamp down to 31, it came out okay, so we went down to 31.

We had the safety lamp on a piece of rope about ten feet long, and we'd lower the safety lamp down over the edge of the skip, bring it back up, if it was lit, we'd bring the skip down about ten feet. And then just keep goin' ten feet at a time. We had a shovel, we'd lean over the edge of the skip and use the shovel to hook the timber so they'd fall all the way down. We got down pretty close. See, what we had to do was get all that timber out so we could run the shaft to check for gas.

There was one timber left, and my partner leans over, tryin' to hook it with the shovel. We knew we were right at the gas. We'd lowered the lamp and it came back out. But it went down maybe five feet or so before it'd go out. So we had went down that five feet, because we wanted to get that one last damn timber outta there. But we just couldn't do it. He leaned over and took a breath, says, "Holy shit, there's gas. Let's get outta here." And I said, "Wait, let me try." So I leaned over and I tried to hook that timber, I got the gas too. You can't see the gas and you can't smell it. But when you'd breathe it in, it'd about knock ya out. I said, "Get us outta here."

So he could ring an 8-1-3-1 to get to 31' or 1-2-3-1 to get to the collar. Well, he started ringin' an 8-1-3-1, then he decided, well, that's going to take too long, so he stopped and started ringin' 1-2-3-1. And so anyway, the hoistman pulled us out, which he shouldn't have because he didn't get a clear signal. Everybody was kinda spooked, you know, it was weird since they got killed. So, he just got us outta there; he thought we were in trouble. Which we were. So we got up to the surface and he came outta the hoisthouse, says, "You guys alright?" I said, "Yeah, oh yeah, we're okay." He said, "Yeah, I got a whole bunch of jumbled bells, I thought I'd just pull ya outta there." I said, "Thank you."

Some mines, some areas, the gas is diluted. It's heavier than air, so it comes in through the bottom, so you keep your lamp on the floor. And there have been times workin' in it, and you get real light headed, and it looks like everything is goin' in slow motion. And then in the case of the Ajax, the gas was so heavy, just like that. It was really concentrated back there. We had to ventilate it constantly; a two foot vent pipe goin' down there with a big fan on 31, blew down to 33'. After they had the shaft pumped out, Paul and A.J. were goin' down to get the timber out. But they didn't run the shaft, didn't let it air out long enough. They got in it and he got half a signal. They were tryin' to ring 1-2-3-1, they got a 1-1-2 out.

The hoistman knew they wanted something?

Yeah, but the law, federal law, if you don't get a clear signal, you don't move the cage or the skip. So, Paul fell out, he went over into the bottom, and A.J. collapsed in the skip.

Back then, there was no organized mine rescue team. That's the only situation like that I've ever been in that somebody got killed. The guys that were there were tryin' to get them out as quick as they can. I think it was Pat Freeman and Gary Owens went into the Carlton Tunnel. Because they still couldn't move the skip. So they went through the Carlton, found out what happened. Like I say, I was off that weekend. I don't know if mine inspectors came in and got the bodies or how they got them out. Once they climbed up to the skip, found A.J. in there, then they could use the skip. So they were able to get out, so I would assume once they got there and found A.J., they just got in the skip and came up and went down after Paul.

What's the story on your nickname?

Somebody started callin' me Jinx. I don't know, that was after I got buried.

I've heard you called Jinx, but instead you sound very lucky?

Oh, God. I was retimberin' the Mary Nevin. It was all caved in with the old crib timber, a lot of muck and everything. Rock, dirt, timber, all froze together. And we'd been workin' down about 200 feet; the first level was 270. We'd been workin' through this cave-in from the surface. We finally got through, oh, the plug was 10, 15 feet thick, and I pulled a timber with that little tugger up there. But when it came out, a hole opened up, so I could finally see the shaft. I had a safety line on, and I was tryin' to see the 270 level. So I start climbing, and I can't believe I did this, I started climbin' down through the timber and I couldn't get down through far enough to see the level, so I undid my safety line to climb down about another five feet 'til I could see the level. Then climbed back up, got the safety line hooked up. When I think back, I can't believe I took off that safety line and climbed down through all that timber. Stupid. I just wanted to see that level, finally. We'd been workin' that shaft all this time, all this time, finally it opened and I could see the level. I was gonna get a look.

Did you work stopes?

Yeah. We had a big stope at the Mary Nevin. We called it the Suicide Stope. When we first got in there to mine, they had a big stope. And the old ladders and timber and everything was all rotten. The Suicide Stope.

But eventually I got back in there and drilled some samples, and they were stoping on a real nice fluorite vein. We got all the old stuff out, then we had to put new stuff in. We had to go up about 100 feet before we could even start minin'. So we had a manway here, it was all cribbed in, enclosed. Then we had a timber slide next to it. Had a little air-tugger, you put in your stopers, your timber, whatever, run it up this slide to your diggins.

Anyway, once you get up to the top put in, oh, I used to put in two stulls, and lagging on top of that to drill off of. When you were ready to shoot, you pulled the timber out, but leave the stulls in. We'd pull the lagging out. Then when we'd shoot, they'd fall down next to the manway where the existing old stope was, where we'd fill the chutes. So as you shoot, you gotta go back up, bar down real good, and put your timber back in. Hopefully your old stulls didn't get shot out, otherwise you had to replace them, or you'd just keep going.

Did you see much high-grading?

Oh, well, I've done my fair share of it. I'd get, you know, samples, just to have around the house and show people. I never went into business, get it to sell to somebody. My own personal collection.

Is it only high-grading if you sell it?

Oh, if you're taking ore out of the mine and the Company isn't getting it, you're high-grading.

D you think it matters if you are making a profit or not?

Well, that would have to be "Mercenary High-Grading," I guess. High-grading for a profit, not for your personal collection. I remember a couple of times I'd find a real nice rock and I'd take it in the office and ask if I could have it. I've heard stories that there was a lot of high-grading goin' on. I mean, all the mill hands, everybody was in on it. You know, you get it out of the hole. Whatever they didn't get down there, it went through a crusher up along a conveyor belt, to an ore bin. So all along the way, you know, the top hand, I heard, he'd get some. Then the guy runnin' the crusher'd get some. Somebody linin' the conveyor belt would see something nice go by and grab it. I didn't know it at the time, but there was a lot of high-grading going on. When you're down in the hole, bustin' your ass and you find some good ore, you don't care who you're working for. All you care is there's a nice rock, and I want it.

Were you ever afraid of underground?

No, I got buried on a Thursday. The Mine was shut down on a Friday, the mine inspector was there to do an investigation. Monday, I was back to work. In fact, I had to go back up into that chute where I was buried and take out the timber they'd used to get me out, so that we could start pullin' chute. That was not scary, but pretty weird, to think a couple days ago I was down there under six feet of rock. Anyway, a safety line on this time, had somebody with me. Got the timber out, and started pullin' chute. My hardhat and mine belt were still there (taking down hardhat and mine belt from where they're hanging on the wall. The hardhat is chipped, cracked and wired together in several places).

Did this largest crack in your hardhat happen when you got buried?

Oh, a big rock came down and got me, another time. Oh, I've been hit by rocks. This big crack here, climbing up in the stope of the Mary Nevin, and the bulkhead we had in there, when we shot, the bulkhead gave way. A whole bunch of rock went down into the manway, and I had to climb up there to get that rock out, it was pluggin' up the manway. While I was goin' up, a rock came loose. I don't know how big it was, I just heard it, and it hit me right on the front of the hardhat and cracked it. One thing or another, you get banged up.

Cresson Mine Collapse

High-Grading vs. Pocket Rock

"Every miner high-grades. Some people do it for different reasons. My reason is my status symbol of the mines I've been in." ~ Guido Honeycutt

Date of birth is June 24th, 1957. We moved to Victor in 1981, for the Ajax mine.

Had you been in mining before that?

Yeah, my whole life. I was fortunate, I started when I was 17 up in Craig, Colorado. Yeah, between my school years. I wasn't good enough to be a farm boy, so I didn't get to go on the wheat harvests. Which I wanted to, you know. I was livin' in rural Colorado, where my family still lives in a town called Cherew. It's over by La Junta. It's a farming community. It means Sparkling Waters, an Indian word.

My first experience was a drill rig, and it was fun. I mean, I was a kid and I was away from home. So I got to see, to be a kid and work hard every day. I stayed in the drilling industry, but moved to New Mexico, in a town called Grants. It's one of the largest uranium deposits in the whole world. I worked out in the uranium field and we did ventilation shafts for mines. We could do a 1000 feet easy, out in that area. Traveled around with them, they were a company out of Wyoming. Teton, they were called. And that's all we did was ventilation.

You would drill into the mine with your rig, and then you would go into the mine and hook up humongous bits. Our rig would do anywhere up to an 8' hole, in diameter. We had rigs that did world record holes, and stuff like that out in New York. I never worked in New York, the deepest we went was about 3500 hundred feet, which is a fairly deep hole. We did a four foot bore comin' back, and they would use it either for reverse ventilation, or sucking in the hole. If we lined it all the way to the mine it was an escape route. We would put a man cage in that would have a cable all the way to the surface, and if you had to, you could get in there one man at a time, and shoot out this hole. You know, on a cable. Pulled you up like an elevator, really, except you're one guy.

What did you line the hole with?

With iron. You know, steel. This rig that we were on would do both phases of the job. We'd drill the hole, secure the top with some kind of concrete pad, and then line the shaft. You know, especially in New Mexico. New Mexico, the whole things got lined all the time, because there was sugar sand out there. In Missouri, we worked in the lead mines out there, called the Galena, which is a lead and sulfur based mine, and these mines are huge. They had big trucks underground that ran on roads, graders that graded underground, full mechanic shops. You know, they were mines. They called the boss wagon in those mines a Chevy pickup truck. You know, where like the Ajax the boss wagon was a little locomotive and you hop in an ore car with it. You know what I mean? A big difference. These were huge mines. You almost couldn't see across one side of the road to another, they're so large. Big trucks could pass underground. Yeah, that was pretty neat mining there. That was hard rock mining, so all you had to do was case the surface of the ground, where what they call the overburden, the decomposed or broken up rock was. As soon as it became hard rock, which Missouri is hard rock like it is here in the Cripple Creek District, so you only had to put 50' in an 800' shaft, and the rest was all solid rock. Unlike New Mexico, with sugar sand.

While we were regulated by MSHA, just like all of the underground mines, we had the best of both worlds. Half the job took place above ground, and half the job took place underground. We weren't true moles.

When is the first time you were underground?

1979. Worked for a company, used to work in ventilation shafts. Had to go underground and work on the part you hook up inside the mine. Which in the small mines it went down in pieces. Like an 8' bore could take five pieces to put together underground, because the mines weren't that big. They were like 9' drifts, or somethin' like that, kind of like what's up here. You know, small drifts. You would go underground and have to hook up a 3' bit, and drill for ten feet, set it back in the ground, and hook up the next stage underneath it. It all bolted together. So you'd hook up the next two foot raise, or whatever, so it became a three to a five, then you'd raise that up inside of there until it got big enough to hook up an eight foot section. Then you'd have a bit that stood 13 foot tall. You know, underground.

The first time I went underground, oh, I was scared to death. I was scared to death. My buddy and I landed this job, and you know, we're pretty young kids. I was 19 when that happened. I was underground, but he couldn't go underground because he wasn't old enough yet. So he had to stay on the surface, and I got to go underground for the first time. It's in a sugar sand mine, like I say, in New Mexico. There was this engineer there, it was his first time underground, too. He was as scared as I was. And when you say scared, it's not like you're not willing to go do it. You're just timid. You're looking at everything, and everything's new. You know, you have a candle in front of you for a light, tryin' to shine around you and look, while your eyes are bigger than anything goin' on. I mean seriously, you do. I was scared, anyways. Part of the ground, back where our bit was, they opened it up bigger. They had it bolted but they had this chemical called Mine Guard, you could spray it on and it adheres to the wire and to the ground at the same time, so the ground can't sluff through the wire. While half the work was done, the whole thing wasn't done because our job wasn't done yet. So you're workin' in a condition that's safe, but not a hundred percent yet.

Then part of the ground, basically, just slipped, or sluffed. I was in front of the engineer for about the first thirty or forty feet, then he passed me. We were runnin'. We took off. You know, it scared us to death. It was a big sluff; it was prob'ly a tons worth of ore. You know, moved down the side, "Wooooosh" this kind of deal goin' on. It scared the crap outta both of us, to be honest with ya. And of course they hollered, and we came back. We were the laughing stocks. Every time you go underground you get a little more educated.

You came here because you heard about work at the Ajax?

Well, yeah. I was out in Missouri, to be honest with ya. Out in Missouri; got married to my wife, Sandy. This is 1980. I got run off a job, which is just part of minin', you know. I was a tramp miner, basically. You went where the money was, and you only listened to so much, and when it was time to go, you knew it. And it was time for me to go, and I left. We were drivin' through Okalahoma City, headed back to Mom and Dad's ranch over by La Junta.

So we were just gonna, we had enough money to go there and hang out for part of the summer and find some work, do the phone call trick. You know, you call your buddies and sooner or later you have a job. It might not be the one you want, but it's a job. Comin' through Okalahoma City, at night, there was derricks everywhere. They just lit the sky up. I told Sandy, "Boy, if I can't work here, where can ya work?" You know. So we stopped, and two days later I had a job in Okalahoma City, workin' in the oil field. Oh yeah, it was a great experience. We made way too much money, and just could not spend everything we were makin', and we were kids then. Young, no kids, and we were tryin' to spend everything. We made enough money where Sandy's checks, we never cashed. So we were savin'.

We told the people, "You know what, on July 4th, we're in the Uncompahgre Forest in Colorado, and we're not comin' back to Oklahoma because there's nothin' here but work." The country out there, you hate to say it, but it sucks. It really does. I don't know if you've ever been to Okalahoma. Oh, it's barren property. It's good for oil, and I guess that's it. There's a lot of farmin' out there, but it's dry land, not a lot of water. They have to pump everything they want, you know. It's pitiful country. Look where we live. Be honest with ya, you know, we live in a pretty nice spot. I had a great time out in the oil field. Learned a new trade while I was there. While it's not minin', for me it's the identical trade I knew. It was a drill rig, you know, and I was a drillman. So it was great.

Nine months later we left Okalahoma, had enough money to take the summer off. Kick back, do the Uncompahgre thing with our friends from Telluride. We came back to Mom's place; we had that trailer that we were travelin' in, we were livin' in. We were almost out of Warranty so we took it to Colorado Springs, dropped it off. I told Sandy, "Let's go up to Cripple Creek." Because I used to come up here as a kid. When we lived at Fort Carson. Matter of fact, we lived at Fort Carson when the Corps of Engineers was re-doing Cripple Creek's dams. My dad got a key to go up there, because he was one of the big-wigs on the post, to be honest with ya. My dad, my brother, my mom and me and my little sister, we used to all go to the Cripple Creek Reservoirs a lot of weekends. Nobody'd be up there, because it was closed for the Corps of Engineers to do the work. We used to just love it.

So we dropped our trailer off in Colorado Springs and came up here with a tent to go campin'. Camped at Bill Buchanan's campground, was talkin' to him, and tellin' him "Oh yeah, we're miners, and yadda, yadda, yadda." You know. "Well, you oughta go to Victor then." "Where's Victor?" You know, boom. Go over to Victor, see what's happenin' over there. Stop by the mill, get applications. Was in Cripple Creek, where the Market used to be. Here's one of my buddies that I left in Missouri, nine, ten months go, okay? I holler at him, and damn sure, it is him. I ask him, "Well, what the hell you doin' here?" "Well, we're drillin' a shaft at the Ajax." "No kiddin' I'm tryin' to get work there, buggin' this guy we called the "Limey." He was this English guy, and he was like under Ed Hunter, like a low-lying Super. He was like a head of all the little Superintendents that worked under them. And I was tryin' to get a job with him, and couldn't get one, and couldn't get one. You know, I'm not a local boy, and don't deserve it. But, I'm a tramp miner, and I think, you know, I'm tellin' him, "I am a miner, I need this job." Day after day, for three or four days.

So we run into my buddy, Dave, who says, "Well, we need help on this job. Why don't you come to work for us again?" I was like, nah, you know, I don't think so. The last one was bad enough. This guy named Rick Sidwell, he's one of our great friends; we worked for three or four different companies together. He was always my boss, we always had conflict, but we were always friends, and he's still my friend today. He came to my house the next day and said, "Let's go. We're goin' to work." I said, "Nah." He said, "Let's go to breakfast, anyway." Over breakfast I ended up gettin' a job with him up there. I told Mr. Hunter that I'm going to work this job, so you can see who I am. Because I am a miner. I know I'm a small guy, but I'm a miner.

We did our two shafts underground. We did some shafts from 3100 level down to the 3200 level, at a very steep degree. 72 degree angle. They were ore chutes, so they could bring the ore off 31, dump it down to 32, and 32 was the shipping station. They could stockpile it up, I don't know, 50 ton of ore, something like that there, and ship it out when they had enough to ship. They shipped from the 32 level, that's where they filled the skips up. Sent them to the surface.

All the miners were pissed at us, because this was a big contract that they didn't get. We went in there and basically did a bang-up job. They came out perfect. It was a perfect job, and the miners were happy after it was done. You know, they were planning how fast they can do it, but we were running machinery and they were running man power. We set up this little machine that would fit this room right here. Just an itty, bitty hydraulic machine, and it did 2 four foot shafts on a 70 degree angle, side by side. Hundred and twenty-five feet deep, or something like that.

So that got through, and Mr. Hunter told me go with this company and when a job comes open, I'll give it to you. And the rest is history. I went to Durango, worked on this machine, kept callin' Mr. Hunter, you know, "I need this job. When's it happening?" "Well, not yet. Call me in a week," and I'd call him in a week. It took about three months, I was gone about three months and we came back. We lived in our 5th Wheel one full winter, froze up, did all the nasty things. That was '81. Never turned back, even though for our type of mining, it's not here anymore. But there's a new mining going on, which in 100 years somebody will be talkin' about that, saying "What do you think of the New mining?" "Well, it's not like us old timers."

I always like to tell people if you want to see what mining is and don't want to go underground, go to the District Museum and get up on that glass ledger that looks down the Portland. Oh, it's way too cool. When you get underground you can take that thought with ya, and you got to where we mined in the Ajax, which is from the 31 down. We actually did a little bit up on the 3300 level, but it was only because they couldn't find the ore they wanted, and we were up there lookin'. But the veins lay at an angle. When you go up inside of the stopes and stuff, you know, you see just how it's laying. It's just like that picture of the Portland. The Ajax is exactly the same. The same veins run though each other. That's what's neat about this mine, are the tunnels that connect. I didn't get to experiment like some of the miners. In the '70s the Ajax had a strong kick that happened before the one that we had, you know, where they re-did the shaft, stuff like that, gearin' up. Where they spent all the money before any ore came out. Those guys talked about goin' back further than what I got to go underground. There's a lot of neat things in this District.

Where did you get to go?

The Carlton Tunnel. Everybody got to do the Carlton Tunnel. I don't know if you heard stories about goin' on the other side of the air doors of the Carlton Tunnel. You could sit back there, if the air doors weren't being used for air passage, you could sit back there. Otherwise, the fan was about six foot tall. It blew the air, either sucked it in, or blew it out, one of the two. You try to fight the barometric pressure you know, so during the winter you try to exhaust out, and during the summer exhaust in. You could shut off the fan back there, and turn off your light, and sit and wait and wait. From that point, the Carlton Tunnel's about five and a half miles away. It's total darkness, except for the light you're carryin'. You just sit there and wait and wait and wait. And all the sudden, it's just like lookin' at those pictures, those 'magic eye' pictures that boom! All the sudden you're in to the picture, you know. It's that same way. Way out there is that dot of light of the Carlton Tunnel. That's five and a half miles away. That's a straight tunnel.

We'd been to there, and up and down that quite a few times. The Carlton Tunnel was never anything but work. You had to go to the Carlton Tunnel, you knew you were shoveling at some point. So you weren't havin' fun. If you got lucky, you only had to shovel either comin' in or goin' out. They had high speed loccies (locomotives) that'd do about 30 miles an hour in that tunnel. There were sand traps where everything was kinda decomposed, the waters course was comin' through, whatever it might have been, that's where all the sand would settle. So you had to remove the sand from the tracks, because the natural movement of the water kept the sand traps right there. You could go through, and if you were only thirty minutes, an hour on the other side, chances are if you went slow enough, you'd go right back through where you just shoveled. Any longer than that, you were shovelin' again. They did put a grate on the front that would try to push it, but thirty minutes of work was way better than three hours of puttin' a locomotive back on the track. Especially, you could step off and Boom! Be up in water to your waist.

The Carlton Tunnel where we are now, if you're sittin' on a Locci, you're probably sittin' about this high off the ground, which would be about four foot. Some places you'd still have to duck, to get underneath it. And if you look at pictures when the Carlton Tunnel was done, it's a twelve foot bore. It's a big tunnel. It might be fifteen foot bore. I mean, I don't know the actual size of it, but it's a huge tunnel. Now with all the sand and silt and stuff like that, I don't know, but I was told that instead of tryin' to fight the sand they were just takin' and puttin' out ties, right on top of the railroad gauge that was below them. Puttin' a new set of tracks all the way out again, right on top of the other ones. I heard there was three sets. If that's true, it's about the height that we're at now. Instead of fightin' this sand, which is a fight every time they went down there. It's the diversion for all of the water, for all of the tunnels around our District. All the water goes through it.

The air doors weren't always open?

If they weren't open, you just opened them. You only opened one door at a time. So what you'd do is open the door, the door was big enough for a Locci and four, five ore cars. The big loccies with four or five big ore cars. The ones that worked in the Ajax were small, they were ton ore cars. These runnin' up and down the track were like three ton cars. So you could open it up, you could pull inside of it, and you'd shut this door. Then you had to walk all the way to the end of it, to the next air door, and open it up.

If you're walkin' in a group of people, like somethin' happened and you had to walk back to the Ajax, or if you're workin' on the other side, I think the tunnel's called the Mohigan on the other side, but I'm not positive of that. But if you're workin' over there, like we core drilled on the other side, long holed and that sort of stuff, but you could turn off your light. You could bob your head like you were walkin' behind your partners, then you shut your light off and they wouldn't notice it. You'd sit right by the air door and when they were about twenty feet from the other air door, you could start pullin' this other door open, and the suction that came in from the other door would cause a water spout to spit up. You could drench everybody, you know what I mean. It's a big fan, moved a lot of air. Screamin' noise, you'd put somethin' in your ears when you went through that area.

Did you always wear ear plugs underground?

You know what, MSHA was a very strict company, that inspected us. They're not like OSHA, who waits for somebody to die. MSHA is on your butt all the time, and I kinda liked it. Because now in my adult life, I wear safety glasses all the time. If I need ear protection I have it, I wear it all the time. Back then you nagged about it, "Oh, I don't need this." But you always had it available to ya, and it was required of you to wear all this safety gear. And to be honest with ya, I have no problems with like safety glasses, I wear them all the time. I look up and if I don't have them on it's like a panic goes off. You know what I mean. They're sittin' on my dash, and I have a spare pair with me, because they're that important. This is most of your work, lookin' up, so MSHA, we had all the safety gear available to us. It was never a problem. It was just if you wore it or you didn't. If you were runnin' a machine, you had ear plugs in. If you were walkin' up and down the drifts, you just didn't have 'em in. You had them in your breast pocket or somethin' like that. You grabbed new pairs every morning and stuff 'em in, and off you went.

What did you think the first time you went underground at the Ajax?

Oh, I loved it. The Ajax is one of the fondest memories, the shortest and fondest memories of minin'. It was the group of guys. It was not bein' a tramp miner anymore for me, which meant travelin' everywhere. The states I worked in were Missouri, Wyoming, Utah, New Mexico, Colorado. That was my area of minin' for eight years, you know, doin' this ventilating drilling, and travelin' all the time. Had a blast. This was my first chance to settle down. Had a company offer me money to buy a house. You know, I loved it. Guaranteed me eight years of work; got to know the guys personally, you know, instead of sayin' "Where's your next job at?" or "How long you here?" Everybody was there to do the job. We probably didn't make them any money, we lost money obviously, because obviously they're not runnn' anymore underground.

We employed a lot of different techniques. Had a German one time come there and show us how to build a real chute. Fred was his name. The Nazi, we called him. He was actually in Hitler's army, not wantin' to be, but forced to be. He was an old timer, and he taught us couple miners a lot of things. When I first went to the Ajax, I worked there as a driller, an underground core driller when I got hired on. And we set our core drill up, it was the least needed core, because we were searchin' for something. We were just drilling small samples, and if we found somethin' then a real core drill would come in.

We ran what they call, like a Winkie. It was on the main thoroughfare between the Ajax and the Portland. The Portland was a storage station for things you didn't use a lot, but you needed underground. So every time the train would come, we'd have to tear down our machine. Go and help them load their stuff up. We were their nippers, is what we were, basically. Whenever miners needed somethin' nipped, I had to do it. I complained a lot, but it was a good complaint. Then when they got through, they'd come back through, we'd set back up. Very seldom did we get any drillin' done. If we got 30 feet in a day, we were doin' good, because we didn't do a lot of drillin'.

When they shut that rig down, and I think they actually shut all exploration down, and they narrowed the crews down, I got to stay at the Ajax, as a gopher helpin' the miners do whatever. I shipped ore, I shipped from that 32 station, I told you we drilled those shafts to the 33 Station. Bill and I one night skipped an ungodly amount, you know, and just had a blast doin' it, when everything works right. There was actually only three of us in the mine at the time, and because there was such small crew, there was three guys per shift and they only had one shifter runnin' back and forth to what they called the C-7, which was the Cresson Mine, which was a horizontal shaft there. A bore that went underground, and it kind of circled around the shaft of the ore body that's supposed to be there. Well, we have Cresson Valley up here now, you know. Which is gone. It's mined; that's the good thing, I guess.

They had a lay-off and I always used to help Gerry Breitenfeld as an electrician. You know, he'd go down a level, call me up, tell me what to do, and I'd help him out. When Hecla came in we all got laid off, and they were regroupin', and they were hiring people back. Everybody they hired back was goin' down to the tunnel, the Carlton, to dig. In those sand pits, okay? Because there was no work for them yet, they were tryin' to get all their miners regrouped and they were hirin' the most essential miners. They'd just hired Gene Leaf a couple of days before I seen Nabil Sawares, he was our Human Resources guy, and I was paintin' the Imperial Hotel for the Mackins. He says, "Oh, they want to talk to you over there." I thought, oh shit, I don't want to go back there, not in the summer and he says, "Nah, you're going to like the job they're going to offer you." Because I was a gopher the whole time, I did all the nasty work. And there was no way I could keep up with Gene Leaf. Gene is a mad machine when it comes to muckin'. I thought I was gonna have to partner up with him.

I get there to the thing and they offer me an electrical job with Gerry Breitenfeld. I tell you what, that was the best day of my life. I went from a gopher to an electrician overnight. Went straight to Zeke's Place, you know, started bragging about it. "Yesterday couldn't spell it, today I are one." You know. "I'm an electrician, guys." (laughter) And I didn't know anything. All I know is I helped Gerry, and Gerry says that's what he needs is a helper. He just needs help, and Guido would be the best one. It was like, too cool. And the rest is history for electrical work for me. So that's like 1983 is when I got started in the trade I'm in now.

It changed things for me. I stayed in the mine, as a mine electrician for about 12 years. I did, well, it wasn't considered a mine, it was done for a City. It was the water tunnel that went from the Air Academy to Rampart Reservoir. Matter of fact, some of the local miners up here, when we first got laid off again, went down to Rampart and sunk a 1000 foot shaft. Well, like a year later they stuck this tunnel down at the Air Academy, that's gonna hit this shaft. 17,000 foot long tunnel we bored. We bored it, yeah. With a tunnel boring machine. Which they don't call themselves miners, they call themselves tunnel rats, and this is the same miner that you see in New York City doin' subways and stuff like that. That's basically what we did, was like a subway.

We had this big machine that bored this tunnel, like I say, 17,000 foot away. This tunnel boring machine was from the '50s. This is a trade, just like mining's a trade, and they have tunnel rats that go from tunnel to tunnel to tunnel lookin' for the big job. We had these guys that would come from Alaska, out to this tunnel. Stopped and waitin' for this job to start in New York, because they knew the boss runnin' it. They worked this machine on different mines, or tunnels, around. You know, they don't consider it a mine, but it's mining, I don't care what you say. It's minin'.

So anyway, we did this tunnel and I was an electrician on that job; got to see it completed. Went from that job to doin' other mining construction. Worked up in Fairplay, in that mine up there. The London Mine. It's on the backside of Mosquito Gulch. On the Alma side of Mosquito Gulch, which Leadville's on the other side of Mosquito Gulch. Talk about a cold job. Work up there in the winter.

We started off up there, me and my buddy went up there to go drillin'. This was when I was an electrician for one year, and the Ajax told us, you know what, we're shuttin' down. Find a job, ya got two weeks and we're done. Hecla. It happened that quick. They called us in one day and said, "Hecla has a policy that we give notice before we start layin' anybody off. We're not at that point, but we're damn close." It wasn't two weeks later we got called in again and they said, "We're there. We're shutting it down. We're layin' everybody off, and we're startin' now. If you can find a job, here's where jobs are hangin' out at." We saw the Fairplay job, my buddy and I. We just had brand new babies; Heidi was just born, and his daughter was just born, and no work all the sudden. And no chance. This was the last big hurrah and it didn't happen. We panicked and ran to Fairplay and got some work. You just thanked God, Victor wasn't that far away.

Did high-grading go on at the Ajax?

(laughter) You know what, high-grading is a bad, bad term. I have a rock from every mine I've ever been in, and I'll have a rock from any mine I go in. And if that's high-grading, I guess yes, it went on. We called it pocket rocks. High-gradin' to me is takin' more than your own collection. You know, where you're sellin' your rock. You're profiting from it. That's high-grading to me. I'm sure it went on. I don't know, I didn't see it. Did I see pocket rock goin' on? You bet. Do I have mine? You bet. And I have my share. This is a Telluride mine. You gotta understand, this is one of a few areas in the world that produces this type of ore. You know, these tellurides. This is the only one goin' in the world, if I'm not mistaken. You know, when the Ajax was goin', I'm sure that's what it was. I don't know if you find it anywhere in the world right now, but here.

You know, when you show a piece to a geologist, which we all know geologists from minin', and that's a nice connection to have with geologists. They all go ape shit over the stuff. And you're talkin' about a rock, that's "Oh, if you like this, you oughta see my pocket rock." You know, but here you need a piece of this, though. And every friend of mine has a piece of this rock. But I did not high-grade it, it's pocket rock. There's a big difference.

Let me tell ya a funny story about high gradin'. God, what company was it? It might have been the original Cripple Creek Mining Company that came in here after Hecla, that bought Bill Reid's spot up here. NERCO. They couldn't hire local people, because we're bad guys. Because we high-grade, okay. That was the rumor, I don't know if that was the truth or not. No word from NERCO, they never said that to me. But there wasn't a lot of local guys workin' up there. And Bill Clymer one day took some ore in there and set it on one of the guys desk and he says, "You know what, I'm a local guy and here's some of your high-grade that I just bought from one of your drillers up top there." And walked out. Every miner high-grades. Some people do it for different reasons. My reason is my status symbol of the mines I've been in.

Did you do any actual 'mining?'

When I was a nipper, the mining that I got to do was when the miners got tired, and I had to drill, like lifters. Or you have to drill a high back hole. I helped them load, I'd get their powder. Matter of fact, I worked for a guy named Brock Gay, you remember Brock? Yes? And Chip Huffman. They were miners when it was the three of us in the mine. I'd go over there and help them set up. That's what the nipper does. You know, it's all goof and no glory, basically. You'd help them set up, and then they would start their drillin', I would go out, I would get the powder, and the igniters, and this sort of stuff, and I'd head back in.

About the time it would take me to do that stuff, or if I had time, I would ship ore and then I would go back in with my powder and my igniters, and nip all that stuff to the front, so it'd be, you know, from here to the wall, 20' behind them. It'd be ready for them. There's where your MSHA violations. You can't have powder hangin' out. But you know, you did these sort of things. You broke the law; did I have my safety glasses on then? No, but most of the time I had them on.

High-grading, or pocket rock. Depends on what you want to call it. I always wanted to be a miner, you know. The money was there. They got paid 10 bucks an hour more than I did. That was my biggest bitch or whatever in the mine. "They make the money, I want to be a miner." So, Brock or Chip would let me run their machine for awhile, while they took a break or whatever. I'd finish off whatever part of their round they didn't have done. Then I would have to start nippin' their machines, and pullin' their lines. While they didn't make me do it exclusively, if they were done they would break their air hoses off and throw the machine over their shoulders. These are both big boys, and you know I'm still a small guy myself. The machine's a struggle for me to carry. But these guys were both pretty prime men at the time, and they carried the machines. I'd nip the hoses, or pull the vent line back, or unhook the last water line so we could do the blast. Or whatever the case may be. Then we'd go load the holes, and we'd blast.

They used to ride about me. They rode about a lot of people. Anyone that whined or bitched. They called me a Whiner, and they'd say, "Cry Baby, Honeycutt. You'll get a raise some day, but no more money." And that's what you called the mine, was the raise, you know. I was like, okay, I guess I deserved that. I'll tell you what, when I became an electrician, the jokes became even harder. They'd be walkin' around carryin' light bulbs, "Who am I? Who am I? Oh yeah, I'm Guido followin' Gerry." Because I wasn't an electrician, I just became his helper. And I went in there and told 'em, "You know what? It's over. No more minin' for me. I'll never do your work anymore. I ARE an electrician now." So it was funny.

The hottest pocket I got to see was on 33 hundred, and they claim it ran 35 ounces to the ton vein. But after we followed this vein, it's the only vein I got to be on that you could see with the naked eye. I mean, you get spots, water courses, stuff like that, but we followed these veins, coming together, until it became one vein, one water course, one slant going up. I mean even the vein is on the slant, when you're looking inside up the water course. We hit a spot that was supposed to be really hot. It was a hot vein, it looked good. We had a chute break around that area, it probably took three or four days to muck it up and haul it up, because I think the Locci was stuck on the uphill side, which there was no exit. It was a pretty funny deal.

It's funny how the different rocks look different. Just goin' over to the Bull Cliffs mines, where the chalk was involved, that's that white streak in the thing, and seein' that ore, in with the chalk.

Were there a lot of pranks underground?

Oh, too many. Too many. Oh yeah. You know what I mean, you always had to look for the next prank. I mean it was never anything that would ever hurt anybody. A couple good examples was greasing the telephones. 'Cause you'd have to come to 3100 level, call on the phone, say everything's clear on 31, we want to run the shaft. Mine rules and they kept everybody alive, is what they did. That was one that was never broken. We had greased the phones. We had two of them. One that was battery powered, and one that was crank, generator type power. I was doin' this pump check with this kid. He gets underground and I tell him to call. Already knowin' the phones are greased, we're settin' him up. Friday night everything got greased, so when the shift came back Monday, you had to watch out for what was greased. He stuck it to his head, and was talking, oh, he cussed and slammed the phone down. He's holdin' the phone out here tellin' the guy, "Call me on the other phone." Because it had to be signaled to run it. So he signals it, picks it up, sticks it to his other ear and the other one's greased. Huh? So he flips out, he doesn't even know how to take it. It was just great. It was like, this is too good, you know. (laughter)

So one time the miners, when I was drillin' in the area between the Ajax and the Portland, the miners would screw with our lunch buckets. We'd go back and eat lunch with them. So we took and nailed theirs down to the bench. We thought, oh man, that is so funny, we got them. Okay. We come back the next day, our lunch buckets are up 20 feet in the air with a big old bolt through it that held rock together. The lunch pail was destroyed. We thought, boy, we better not screw with them anymore.

(laughter) The boys and their toys. We had hammers, and they had mining equipment. They took their mining equipment and stuck our lunch pails way up on the ceiling, and we thought okay, we see where this game's goin', we can't win this one. That's the kind of play that went on, and it went on all the time. Always. It's good. You can take and eat somebody's lunch all the time. That went on quite often. They stuck a chipmunk in somebody's lunch pail one time. They caught it up in the office, stuck it in somebody's lunch pail, so when that chipmunk got underground, it took off. Don't remember what happened to it; think it just went off in to the Carlton. In the Carlton it's over for any animal that doesn't swim.

Were there a lot of critters in the Ajax?

You know, beavers would come up there. That's the most that I saw. I never seen bats or anything like that myself. I didn't see the beaver. This is another funny story. We're stittin' at the Ajax, on the 3100 level, which is the meetin' room, or lunchroom. You got hard rock all around you, a shaft, all kinds of electrical equipment, but you didn't have to wear a hardhat down there, because it was the lunchroom. It was classified as a lunchroom. So everybody's sittin' there, and I think it was Joe Cook came up, I think it was Cook, Stitt, and Gary Owens. Joe comes up and says, "Oh yeah, we ran in to this beaver." He sounded like a pretty respectable beaver, two foot, almost, you know. The next guy, oh, the beaver got a little bit bigger. Gary was probably five minutes behind him. We had a lot of nicknames for Gary. He always had a cup of coffee.

I remember workin' with him one time. God, I hope Mr. Hunter don't read this, but we were workin' down at the, I forget what tunnel it was, it was like off of the 500 level for an escape shaft. Anyways, it was right where the Economic Mill was, in that area there. Gary was a good worker, but if he could beat somebody out of time, he would. 'm workin' with him, and we're doin' this cribbing, goin' in there and he says, "Let's go out and sit in the truck." It's cold down there, and durin' the winter those guys shouldn't have been workin' there, but they were. It was just a place to keep a guy, to be honest with ya, a place to keep him workin'.

We'd eat a sandwich, and he'd call it the "Half-a-Sandwich." He'd say, "Don't eat that sandwich, all of it. Keep half of it." I'd say, "Why?" and he'd say, "Oh, you'll see." And if you were sittin' outside in the truck, when you shouldn't be sittin' outside in the truck warmin' up or somethin' and somebody drove up on ya, he'd take this half a sandwich out and it was called the "Half a Sandwich Deal." Whenever somebody pulled up, he'd say, "Oh yeah, just takin' a quick lunch break here. We'll get right back." Gary talked real fast, he always had everything lined out, too. Loved workin' with him, he's a funny guy. So the half a sandwich would always keep him out of trouble. I tell everyone that story. "Keep that half a sandwich, I'll show you why later."

What about accidents in the mines?

I had a best friend killed in the mines, back when I was like 19 years old. In New Mexico. We were split up, because we were too good of friends. That's the way some bosses think, that friends can't be working partners. You know when you're a kid you always say 'Don't cry over my funeral, drink a beer.' And we did, we got drunk and missed the funeral. He was in our hearts. MSHA has a thing, it's called 'Fatalgrams'. Whenever someone dies in a mine, these Fatalgrams go to every registered mine in the country, maybe in the world, I don't know. In the country anyways. It says what happened, how it can be avoided, and the new rule that's going to make you avoid it.

What happened, there was this mine was in desperate need of air. This was a six foot shaft, 800 foot deep. They pulled the bit out, we used to pull the bit out at night so we could set the rig up, get it ready for the next crew. Because this rig had to be moved forward 8 feet, tilted back to a 30 degree angle, or whatever, to make it to a crane, instead of a drill rig. The drill rig was vertical, and the crane was slanted over, so it could set this casing. He was walking by and tripped, and tried to jump over the hole, because he was falling towards the hole. The hole was what they all a 'sucker.' Some of them blow air out, some of them naturally suck. Depends on the time of year, and what the mine is needin' at the time, what it's going to naturally do. And this was pulling the pebbles off the side, type of sucking. And this is only hours; once the mine equalizes, things change, of course.

He tried to jump the thing, is the story I heard. I wasn't there. Yeah, drug his ass right in and killed him instantly. 800 foot drop; closed casket funeral. Good buddy, miss him a lot. Everything brings a tear to your eye when you think about a mine accident. But they ranged from everything.

Gary Horton got hurt, and that's how I got my job at the Ajax. He got hurt within two weeks of them shuttin' it down, keeping a skeleton crew. I took his job, so I was hired right back. That's when I worked with Chip and Brock. Joe got hurt one time real bad, I don't know if you remember that, when he cracked both shoulder blades, or somethin' like that.

You know, as far as seeing accidents happen, I got one myself. (shows his hand where there is a large lump under the skin). Yeah, that happened. I was 19, my drill bit, the bit I told ya went up in stages, well, sometimes we had to take these bits and set them back in the mine and service 'em. And this bit fell apart so we had to connect a new one. I was connecting it and didn't hear the rocks. You could hear it 'tink, tink' off the pipe. Because you're in a 17" hole with a 10" drill steel, so the hole around the bit's not that big. So any rock, you could hear it doin' that. Well, I didn't hear it, and it put a big bump on my hand. That's when I was 19, that happened. From a rock from how far, who knows. A little rock.

I've seen it first hand, you see guys sweatin' too much. This happened in Missouri, sweatin' too much. This guy totally got that big rock, he was down there tryin' to break this PVC, because you lined your overburden in PVC while you were drillin' it, so that when you came out, you could break it away. When the drill bit was comin' out, because if you go down 17", come out four foot or something, half of it is hits pipe that 17" around. Okay, it's getting pushed out of the ground, so you had to break it away. This guy was down there sweatin' too hard. Missouri, I don't know if you've ever been there, it's beautiful country, humid as hell, and to us arid people out here, where this is no humidity, it's hard to take. So he's sweatin' bad, takes off his safety glasses and swinging this hammer, had a piece chip off and hit his eye. He got lucky, he didn't lose his eye. But he lost about 85% vision. So that was a bummer.

We were doin' this tunnel, the one at the Air Academy to Rampart, and we had these, what you called 'slip forms,' and they were for the concrete, and it was on this machine that the forms butterflied into place, and you used airdraulics. And the guy had his finger in the wrong spot, he was runnin' the controls himself, (chuckling), it took his thumb and pancaked it to oh, the thickness of a quarter, something like that. Oh, I just seen the blood blow out like this. I'm lookin' at this guy thinking, I'd be screaming. He just calmly ran the machine back, looked at me and says, "Well, I think I'd better go to the doctor." It was like, my God, I couldn't believe it. It just squished it so bad, so fast, and it just had to hurt. It was like one of those, I know I would'a been crying. Tough as nails, I guess. He wore a big ol' bandage for months.

And they get ugly. One time we were, I told ya about runnin' the shaft all the time, from 31 to 33. Every Friday, Bill Clymer, Gerry Breitenfeld, who was my boss, and me, we used to go down and service all the equipment down there. Bill did the mechanical stuff, and Gerry and I did the electrical stuff. It was just a good maintenance program is what it was. Gerry had this all set up. So these guys came up to 32, got off the skip. We rode down to 33. About 20 feet below 32, the gas was there. I don't know if you've heard about the gas in these mines up here. Gerry, knowing that it's associated with the smell of wet timber, if it comes out around timber, it permeates this type of smell. He smelled it, and he said, "Gas." And this cage at the Ajax, you could put five in there, tight. Four was just comfortable, and there was three of us in there. These are small cages that we're in. So I grab the dinger, the bell, and I ring the emergency signal. And Gerry knows everything's okay. He was smooth. He says, "Grab Bill." We grab Bill and pull him down, and we're in the bottom of the skip. And we're hyperventilating, because there's no air and we're trying to gasp it. And nothing happens to the skip. Gerry says, "Oh God, he didn't get the signal."

I have to stand back up and I ring 8-1-3-1, which means rise to 3100. And the hoistman was runnin', he apologized and everything, but we're pissed off. He said he knew he screwed up once, he didn't act for the emergency signal. He wasn't sure he heard it. Okay. We all survived. He raised us up to 3100, and what the deal was, is you raise slowly past 3100, and if you get a stop signal, everything's okay, and if you don't, it's a haul ass 3100 feet to the surface and have an ambulance waitin' on ya. That's the rule. But we signaled, we got out. Bill wasn't hurt. It was good we were together. Gerry was the calmest; I got to do the ringin'. So what happened was the fans had shut down inadvertently, because of an electrical storm or something like that. I don't remember what it was. So there was no ventilation from 3200 on down to 3300.

When there's no positive ventilation, the gasses accumulate until they accumulate so thick, they just start risin' like a fog. And by the time all the big bosses get down there from Hecla, because they're safety conscious, "What happened, yadda, yadda." We almost died and everything else, but we're standin' there watchin' this fog come outta the shaft, on the 3100. It was actually about this deep on the floor, about four feet, before it hit the Carlton. It was clean air, then. You could see it rolllin' out and when we got the fans back on everybody had to evacuate that side of the mine. Basically what you had to do was go to the uphill side of the Carlton Tunnel, which would be towards the Portland. Because the natural ventilation's gonna make it turn the corner at the Ajax.

The Ajax came back and 'T'd basically, and that's where the Carlton was. If you went left, you went up to the Portland, and if you went right, you went down the Carlton, and whatever drifts were along the way. So the uphill side was the Portland side, and the downhill side was the flow of the water. You'd just watch this mist, I don't know what you'd call it, it was like a fog. You could see it. You could see it hangin' out. I'm sure it's not the gasses that you're seein', the lack of oxygen that you're seein', you're seein' the mist associated with it. It was like one of those eerie films, you know. You could actually take your light, your flame safety lamp, and actually lower it into it and watch the light go down like this, and bring it back up, it'd go on. Oh yeah. It was neat after that, but it was scary. And I'll tell you what, things roll through your head. I don't know that we were close to death, but we were dyin', you know what I mean?

We were dying if everything hadn't worked out. If Duncan hadn't got the signal the second time and went ahead and lifted us out, we would'a died right there. But you're thinking, Oh God, my baby's not born yet, you know. The next thing you know you're safe, and everything's okay. You're pissed off at the same time. When it comes down to it, it was our fault. These guys came off the 32, they did not come off the 33, where the gas is. There's no gas on 32, it's not a mine in that area. It's a dugout as big as your, three times the size of your kitchen, if it was made into a circle, is all it was, with these two 4' shafts comin' in to it. That's all. And a machine to run. That's all that was there. So there's no place for the gas to be from, you know. So standin' on 33, where the workings were, it was scary stuff. So that's a close incidence, but not death. And this gas has taken many lives in every book you read, there's always somebody that's died from those gasses.

They claim if the gas is heavy enough, it will kill birds on the headframe. I don't know if that's true. Who knows if there's any truth, but it's a damn good story. When the barometer's low is when it's bad. You put your diggers on in the morning, you'd walk in, you'd grab your light, put it on. Grab your brass, brass is part of the system. Brass in. Brass out. Brass, so when your dead they can read your brass; find out who you are. That's basically what it's about. Who's missin', and if you're dead. That's the only things it's for. Where you're at, are you missing, and are you dead. You know. You do all that, you'd walk right over to the barometer and take a look at it. If it was low, you were more cautious. If it was high, it was like, okay. Things are better when it's high.

What was it like the first time you went to work at the Ajax?

That was my first full-time minin' job. Oh, I was experienced, it just wasn't their experience. I was a driller. My machine was just bigger than their machine. It's no difference. They had to man handle their machine, and I had a machine that I controlled. That was the only difference. I controlled my machine, and their machine, they had to drag and carry. Their machine weighed 120 pounds; mine weighed 120 tons. I could talk the best with them. The game's the same, it's just what you're doing. It's like callin' these guys "Tunnel Rats." They're miners. The only difference is they're not goin' after an ore body. That's the only difference. They're just diggin' holes. It's the process of mining, without the glory at the other end.

What do you think of Tommyknockers?

I believe in them. You gotta prove them wrong first. It's like, I live in an old Victorian, I'm sure I've got somethin' happenin' in there, too. We got things that happen that you can explain to everything, the house bein' too tall, the wind blowin, the water drippin' in the mine. Turn off the light sometime, and just sit down there and just listen to dead silence. It's scary, just sittin' there listening. You can't do it very long, you gotta... 'see if my light still works.' Okay. So about five minutes, it's hard to sit down there. But I was goin' back to the Vindicator one time and actually had to go by myself.

Gawd, I'm not supposed to. Always do. Okay, it's just part of the game. I'm going back there, and you know how these lights are. It's a nice beam up front, but it's shadows on the sides; your exactly 180 degrees, or a 190 degrees, off each side is all shadows. It's a lot of this (bobbing his head around) lights going around the rock, doin' this number here, making shadows. When your heart's thumping, you're lookin' all the time. I always told everyone, "Don't jump up on me." Because I always took what they call a 'Buzzy'. I carried a buzzy with me; I held it in my hand. It's a ten inch pipe wrench. It's also called a wrap wrench. It's made for miners, and it's a pipe wrench that opens up to work most common mining stuff, and on the other side it was a hammer. Basically, if you were trapped you were supposed to take Buzzy, you could actually beat on the walls and send out echoes, or beat on the pipe, okay. And that's what this wrench was for, too, besides a work wrench. The bigger the miner you were, the bigger the wrench you carried. They put studs on the end of them. They'd take big ol' bolts and weld them on the end so the handle would be bigger, or different angles, depending what you used it for. Myself, a little wrap wrench, is all I needed, you know. I'm an electrician now (chuckling), I don't need big tools.

But anyways, I'm goin' back to the Vindicator. We were startin' to ventilate it, but where the miners were going to work we had this rig comin' in from Sweden that they wanted to show off. They were goin' to take it to the hard rock at the Ajax, and show it off. We were gettin' power set up for it and this glove floats past me. It's an old glove, okay, and it's about the size of two hands. It's swelled up from time, you know, the old rubber did. And it caught my heart, It's like, "Oh my God." You're mind starts racin'. Mine raced about what if there is something back here that glove fits?

It was just floating down the side in the water. You know, just floating away. It's one of those things, it's like "Oh!" And this is back in that mine, when it becomes a dead mine. The muck on the walls is all slimy and stuff like that. You know what I mean? If it's active with air and everything, that stuff is not there.

That's somethin' else that's neat about goin' back to the Portland is the permafrost. Have you heard of the permafrost back there? You go from the Ajax, and about two-thirds of the way to the Portland, just like a line in the sand, there'd be a ring around there and stalagtites would be all over. Little iridescent mushrooms would start growin' back there. You had to take a jacket with you, if you were going to work in the Portland. The Ajax is 80-some degrees, 100% humidity, whatever it was. It was hot and wet and when you get back there and it's cold and wet. Just right around the shaft, because it sucked almost all the time. Very seldom did it change direction; it was natural. For a third of the distance, which you know the Ajax and the Portlands were apart, a mile or a mile and a half, something like that to the last one. Probably a third of that, 5, 6, 700 feet of it, you know, was bitter cold. Put on a jacket. Put on your gloves. And then you could be standin', looking at the shaft, 30 feet from the shaft, walk another 40 feet this way to the Vindicator "Y", start stripping your clothes off again. Because it all went out towards the Carlton Tunnel. Dead air started on the other side of that, and a ring right where it ends.

There was really frost on it?

Yeah, oh yeah. Freezing cold and with the air blowing there's ice down in there, and I don't know exactly where the permafrost was up in that shaft somewhere. The air that came down brought it with it, and out. It made neat designs, is what it did. And these stalagtites, you could reach up and they're some sort of calcium. You could squish them like that, you know. Some you could break off that were more solid, almost like a crystal type of deal. And the mushrooms, I don't know if you've ever seen a mushroom that grows and never gets light, it's almost translucent. I don't know if it's a real mushroom; it looks like one. It is weird.

When you were walking and you saw the glove, did you think about Tommyknockers?

Oh yeah, of course. It scared me; that's what I was thinking. If somebody's that big back here, and I know the logic side of your brain is telling you that glove is an old glove, it's from the '50s or something. It wasn't that old of a glove; it's not like it was a hundred year old glove. I'm probably exaggerating how big it was, but I could probably fit two hands inside of it. The thumbholes looked to be an inch and a quarter around, so it's just puffed up. By then you have the light off of your hat, so you can control better where you're looking without jerking your head around. (laughter) Next thing ya know, it's comin' from behind ya. And maybe people that don't believe in supernatural stuff don't worry about that sort of thing, but I kind of believe in all that until it's proven not there. Right now all I know is somebody else believes in it besides me. You know. But as far as seeing weird stuff, that never happened.

Did you ever feel a presence underground?

You mean the kind of stuff that goes down the back of your neck? All the time. All the time, yeah. And whether I thought I was feeling a presence, or it was just another shadow that scared me ... as long as I had my wrench out.

Do you think there were ghosts underground?

I don't know if you'd call them ghosts or not. For lack of a better word, yeah. I was just scared, to be honest with ya. In a group it never happens. It never happens when someone else can die with ya, okay, because you could have conversation, or they're walkin' with ya, whatever the case may be. It's always when you're by yourself. The more your mind can wander, the better the victim you are and mine wanders. I'm not chicken of anything, but if you don't know, I mean what could come out? The stories are way too good. If you try to solve it, it's a mathematical problem, it's not even my story then.

Mining is in everybody's heart. I don't care who they are, or how much they complain about it. It's our life. It's everybody's life. If you can't grow it, you have to make it. And if you gotta make it, you have to mine it. Whether it's fossil fuels, or gold for our computers. You know, there are a lot of people that don't like what it represents I go up to the Lookout with my binoculars, and I look down in the pit. I watch what's goin' on. It's a different style of mining.

Vindicator Mine

GLOSSARY OF (*SOME*) MINE & MINERAL TERMS

Assay. 1. To test ore or minerals by chemical or blowpipe examination. To determine the proportion of metals in ores by smelting in the way appropriate to each. Gold and silver require an additional process called cupelling, for the purpose of separating from them the base metals. *See* Fire assay. 2. An examination of a mineral, an ore, or alloy differing from a complete analysis in that it determines only certain ingredients in the substance examined, whereas an analysis determines everything else it contains.

Adit. 1. A nearly horizontal passage from the surface by which a mine is entered and unwatered. In the United States an adit is usually called a tunnel, though the latter, strictly speaking, passes entirely through a hill and is open at both ends. (Raymond) Frequently also called Drift, or Adit Level. 2. As used in the Colorado Statutes it may apply to a cut either open or under cover, or open in part and under cover in part, dependent on the nature of the ground. (Electro-Magnetic Min. & Dev. Co. v. Van Auken, 9 Colo., p. 207; 11 Pacific, p. 80.)

Bonnet. 1. A covering over a mine cage to shield it from objects falling down the shaft. (Raymond)

Cage. A frame with one or more platforms for cars, used in hoisting in a vertical shaft. It is steadied by guides on the sides of the shaft.

Calaverite. A telluride of gold and silver, (Au. Ag) Te_2. Variable in composition, but contains about 39.5 per cent gold and 3.1 per cent silver. (U.S. Geol. Surv.)

Collar. The mouth of a mine shaft.

Dog. Any of various devices for holding, gripping or fastening something.

Drift. A horizontal passage underground. A drift follows the vein, as distinguished from a crosscut, which intersects it, or a level or gallery, which may do either. (Raymond)

Fire assay. The assaying of metallic ores, usually gold and silver, by methods requiring a furnace heat. It commonly involves the process of scorification, cupellation, etc. (Standard)

Gallas frame. Phonetic pronunciation and spelling of Gallows frame in the Cripple Creek - Victor Gold Mining District, and other parts of the American West. *See* Gallows frame. Perhaps begun as an intentional mispronunciation to differentiate those who were local to a District, from those who were not.

Gallows frame (English). The frame supporting a pulley, over which the hoisting rope passes to the engine. (Ihlseng) *See also* Headframe, Gallas frame.

Guides. The timbers at the side of a shaft to steady and guide the cage.

Headframe. A structure erected over a shaft to carry the sheaves over which the cable runs for hoisting the cage. Called in England, Gallows frame. (Chance)

Hoist. 1. An engine for raising ore, rock, coal, etc., from a mine and for lowering and raising men and material. Also called Hoister. 2. The amount of ore, coal, etc., hoisted during a shift.

Krennerite. Orthorhombic telluride of gold and silver. Composition variable, (Au, Ag) Te_2. At Cripple Creek analysis gives gold at 43.86 percent, silver 0.46 per cent; tellurium 55.68 per cent. (U.S. Geol. Surv.)

L-208. Issued by the U.S. War Production Board, from 1941 -1945 Order L-208 effectively ended mining of all materials that were non-essential to the efforts of WWII.

Lateral. A horizontal mine working.

Level. A horizontal passage or drift into, or in a mine. It is customary to work mines by levels at regular intervals in depth, numbered in their order below the adit or drainage level, if there be one. (Raymond)

Lode (Corn.) Strictly a fissure in the country-rock filled with mineral; usually applied to metalliferous lodes. In general miners' usage, a lode, vein, or ledge is a tabular deposit of valuable mineral between definite boundaries. Whether it be a fissure formation or not is not always known, and does not affect the legal title under the United States Federal and local statutes and customs relatives to lodes. But it must not be a placer, i.e., it must consist of quartz or other rock in place, and bearing valuable mineral. (Raymond)

Muck. 1. Earth, including dirt, gravel hardpan and rock, to be, or being excavated; overburden. 2. To excavate or remove muck from. 3. To work hard; to toil. (Webster)

Mucker. (U.S.) One who loads mine cars, and, in most mines, is also a trammer pushing the cars to the shaft, tunnel or adit mouth. (Weed.) 2. One who excavates earth, or muck. (Webster) This term applies more especially to metal mines.

Petzite. Telluride of silver and gold, (Ag, Au)2Te. The gold content ranges from 18.2 to 25.6 per cent and the silver from 40.7 to 46.8 per cent. (U.S. Geol. Surv.)

Raise. An opening, like a shaft, made in the back of a level to reach a level above. (Standard) The term is in general usage at mines in the Western United States.

Shaft. 1. An excavation of limited area compared with its depth, made for finding or mining ore or coal, raising water, ore, rock, or coal, hoisting and lowering men and material, or ventilating underground workings. The term is often specifically applied to approximately vertical shafts, as distinguished from an incline or inclined shaft.

Sheave. A wheel with a grooved circumference over which a rope is turned, either for the transmission of power or for hoisting or hauling (Chance). Any grooved wheel or pulley. (Webster)

Skip. 1. A large hoisting bucket, constructed of boiler plate, which slides between guides in a shaft, the bail usually connecting at or near the bottom of the bucket so that it may be automatically dumped at the surface.

Stope. 1. An excavation from which the ore has been extracted, either above or below a level, in a series of steps. A variation of step. (Standard) Usually applied to highly inclined or vertical veins. 2. To excavate ore in a vein by driving horizontally upon it a series of workings, one immediately over the other, or vice versa. Each horizontal working is called a stope (probably a corruption of step), because when a number of them are in progress, each working face being a little in advance of the next above or below, the whole face under attack assumes the shape of a flight of stairs. When the first stope is begun at a lower corner of the body of ore to be removed, and after it has advanced a convenient distance, the next is commenced above it, and so on the process is called overhand stoping. When the first stope begins at an upper corner, and the succeeding ones are below it, it is underhand stoping. (Raymond)

Stull. 1. The top-piece of a set of mine timbers. 2. A timber prop supporting the roof of a mine opening. (Weed) 3. (Corn.) A platform (stull-covering) laid on timbers (stull-pieces), braced across a working from side to side, to support workmen or to carry ore or waste. (Raymond)

Sylvanite. A gold-silver telluride, (Au, Ag) Te_2, containing gold and silver in the atomic ratio of 1 to 1. This required 24.5 per cent of gold and 13.4 per cent silver. (U.S. Geol. Surv.)

Tellurium. A rare element analogous to sulphur and selenium, occasionally a native crystalline substance of tin-white metallic luster, but usually combined with metals, as with gold and silver in sylvanite. Symbol, Te; atomic weight 127.5; specific gravity 6.27.

Telluride. A compound of tellurium with another element or radical (Webster). Often rich in gold and silver.

Tram. 1. A boxlike wagon, now often of steel, running on a tramway or railway in a mine, for conveying ore. (Webster) 2. To haul or push trams or cars about in a mine. (Gresley)

Tunnel. 1. A tunnel, strictly speaking, is a subterranean passage open at both ends. An *adit*, if continued through a hill, would then be a tunnel. (Skinner) Any level or drift in a mine open at one end, or which may serve for an adit. See Adit. (Century) Often used as a synonym for Adit, Drift, Gallery. 4. To make an opening, or a passageway through or under, as to tunnel a mountain; to cut, blast, or otherwise make a tunnel.

Vein. 1. An occurrence of ore, usually disseminated through a gangue, or veinstone, and having a more or less regular development in length, width and thickness. A vein and a lode are essentially the same thing, the former being rather the scientific, the latter the miners' name for it. (Century)

Winze. A vertical or inclined opening, or excavation, connecting two levels in a mine, differing from a raise only in construction. A *winze* is sunk underhand and a raise is put up overhand. When the connection is completed, and one is standing at the top, the opening is referred to as a winze, and when at the bottom, as a *raise*, or *rise*.

In Part: Department of the Interior Bulletin 95 Franklin K. Lane, Secretary
Bureau of Mines Van. H. Manning, Director
"A Glossary of the Mining and Mineral Industry"
by Albert H. Fay; Washington Government Printing Office, 1920.

In summary....

The hoistman sits in a lofty throne
It'd be smart of you to throw him a bone
The miner drills and then he blasts
While the mucker mucks till it kicks his ass
A nipper fetches what you need
Then clears the ditch where someone peed
Trammers run ore up to the skip
Skippers bring down what nippers will nip
Rock doctors locate the underground gold
High-graders take some, it's already sold
Timbermen shore to keep loose rock at bay
Blasters get a round in and out each day
Underground life is pretty routine
The work is hard, conditions obscene
Without miners there'd be no ore
Watched by world markets, told of in lore
There'd be no silver, there'd be no steel
No millions gained while little hogs squeal!
~ Sylece Andromeda

'Run A'Muck'
Winner "2006 Miner's Poet Laureate"
National Mining Hall of Fame & Museum
Leadville, Colorado

Anchoria Leland Mine